Building Apartheid's Legacy: Black Businesses in South Africa's Construction Sector

Furguson

Table of Contents (Volume 1)

CHAPTER 1

INTRODUCTION

1.1 BACKGROUND TO THE PROBLEM

The construction sector of the South African economy consists of a variety of types of enterprises. These enterprises engage in 'formal' and 'informal' construction and building activities. The distinction between these types of activities is that the latter are not recorded in official returns (Krafchik, 1990:5). Within both the formal and informal groups large-scale enterprises (LSEs), medium-scale enterprises (MSEs) and small-scale enterprises (SSEs) are found. These enterprises are owned by individuals from various race groups.

The majority of formal firms are white-owned and they tend to dominate the construction sector, undertaking most of the large civil projects, housing schemes, township developments and public work. They are highly organised, and tend to form large representative bodies which effectively control the circumstances under which their members operate. There are black firms which operate in the formal sector, but they form a small minority (see section 1.4.1 for definition of 'black'). Low levels of education and training amongst Blacks and the effects of Apartheid have tended to make it difficult for black-owned firms to establish themselves in the formal construction industry. As a result, black-owned firms have tended to operate at lower levels of technical and managerial sophistication than their white counterparts and in different, less lucrative markets. It is evident that there is a large fraternity of black contractors, operating various types of SSEs. These enterprises operate mainly in the black residential areas or find employment as sub-contractors to the larger white-owned developers and contractors. These firms are referred to in this dissertation as 'black-owned small-scale construction enterprises' - or black SSCEs.

South Africa, in the early 1990s, is poised to undergo a fundamental transformation from white minority rule to full democracy. Once this has occurred, it is likely that many foreign states, organisations, and institutions will become actively involved in nurturing and facilitating her economic development. The early signs of this have already been seen in the recent World Bank and USAID missions to the country. It is quite clear from the literature, conferences and action that have flowed from these missions, that the: construction sector; provision of housing and; development of small- and medium-scale enterprises (SMEs) are all regarded as important targets of development strategies.

Krafchik (1990) has argued that the recent trend towards sub-contracting has benefitted LSEs at the expense of SMEs. Internal and external factors constrain these sub-contractors. Krafchik (1990) noted that the internal constraints were reflected in poor management, while the external constraints stemmed from: the low effective demand for housing; State land allocation policies; and the conditions under which the sub-contracting relationship occurs. Krafchik (1990) has further argued that, because of these external constraints, the trend towards sub-contracting has limited potential to advance 'inward industrialisation' (a is, "the potential link between controlled Black urbanisation and economic growth" (Krafchik, 1990:63)), but under different conditions, it could.

As South Africa moves into an era of democratic government and equal opportunity, it is likely that pressure will be brought to bear on the control that Whites effectively have over the contracting industry. It will not be sufficient to agree over future equity of opportunity. It is generally reflected in leftist political views that part of the process of democratisation involves redressing past wrongs. In this regard, black SSCEs have not had equal opportunities that facilitated the training of their owners and their potential to gain access to the formal construction industry. It is thus an important part of the transition to an equal society that the development of these enterprises be supported.

The shortage of housing in South Africa will have to be addressed in the coming years and the first democratically elected government will be under considerable internal and external pressure to prioritise the provision of housing. There should thus be an increase in the effective demand for housing. In addition, it can be expected that existing State land allocation policies would be significantly altered by a new government. This means that two of the external constraints noted by Krafchik (1990) are likely to fall away. Further, it can also be expected that developers will be under closer scrutiny and greater pressure to ensure that sub-contracting relationships are non-exploitive and that community-based sub-contractors are preferentially employed. The scenario constructed by Krafchik (1990) is thus set to alter fundamentally. In future housing developments SSCEs, whether sub-contractors or general contractors, will be afforded opportunities that, historically, they have largely been denied.

Where Krafchik (1990) focused on sub-contractors, this dissertation will broaden the scope to include a range of SSCE types, from sub-contractors to general contractors. As will be elucidated below, the problem area that this research addresses concerns the constraints experienced by black SSCEs and the resultant effect on their capacity to play the role intended for them in a future South Africa. It will be hypothesised that SSCEs do not have the capacity to play this role - at least not without some degree of support and development.

This dissertation thus flows from Krafchik's (1990) work. His arguments are accepted, but with or without these arguments, the new South African government is almost certain to provide or facilitate housing and is also likely to insist that SSCEs are employed in this process (ANC, 1993). Many argue that this will result in important benefits for the whole economy, given the strong linkages between construction and other sectors (Krafchik, 1990; Hendler, 1989; Currie, 1971). Perhaps more importantly, though, it will provide steady access to work opportunities for construction and other SMEs. The damage caused by past state policies cannot easily be undone. A major part of this damage has manifested as skills-, management- and entrepreneurial-training deficiencies (Krafchik's (1990) internal constraints) among Blacks. Those SSCEs currently operating will not be able to obtain such training if it interferes with the day to day operation of their firms. In other words, on-the-job-training is one of the few methods of delivering training to existing SSCEs.

An important opportunity thus lies ahead in that the development of black SSCEs can be coupled with the provision of housing in South Africa. Direct benefits could flow from this - the housing shortage would be diminished, the economy would be stimulated and the need to compensate for the historical suppression of black SSCE development would be addressed.

1.2   THE PROBLEM

1.2.1   The Problem Area

The problem area that this research addresses is the following. While accepting that Krafchik (1990) has correctly identified the internal constraints for his Western Cape case study sub-contractors, it is not clear whether or not his findings are representative of all types of contractors, nor is it obvious whether or not they are transferrable to other regions in the country. In addition, little is known about the effectiveness of those institutions that currently offer support to SSCEs, nor of their capacity or willingness to do more. However, the external constraints identified by Krafchik (1990) are set to change. This establishes a need for the current knowledge on SSCEs to be refreshed and extended.

The following needs to be established from a range of black SSCE types: general details of the firms; details of the owner's personal background; their training and work experiences; what they currently do, how they do it and for whom they work; their business management capacity; their future objectives; and most importantly, what *they* perceive to be the constraints to their further development.

1.2.2   The Problem Statement

The problem that this research addresses is that of identifying the internal and external constraints faced by a range of types of South African black SSCEs, and assessing their capacity to contribute to the provision of housing.

1.2.3   The Importance of the Problem

A number of development and/or support institutions exist which specifically target black SSCEs (see Appendix F). The very existence of these institutions is evidence that the importance of SSCE development is widely acknowledged. In addition, it is clear that the new South African government's economic and housing policies will attach great importance to the stimulation of SMEs (ANC, 1993). It is intended that this study will inform the process of SSCE development by investigating what is needed and suggesting how it could be supported. In the prevailing economic climate, it is imperative that whatever resources are channelled into development are employed in such a manner that the long-term benefits to the construction sector are optimised. It is thus intended that the recommendations of this study would serve as a guide to a institutions concerned with the development of black SSCEs.

1.2.4   The Precise Purpose of Investigating the Problem

The purpose of this study is to investigate the problems that constrain the capacity of black SSCEs to play a significant role in the delivery of formal housing, and to evaluate the existing support/training/development infrastructure in order that recommendations may be made regarding how growth can be promoted.

1.3   THE HYPOTHESIS

That, because of internal and external constraints, and the ineffectiveness of the training/support/ development infrastructure, black SSCEs do not currently have the capacity to play a significant role in formal housing development.

1.4   SCOPE AND LIMITATIONS

1.4.1   Sample Group

The sample group has been confined to 'black' contractors. In this dissertation the term has been taken to mean 'African' contractors (of a race group other than 'white', 'coloured' or 'indian'). The reason for this limitation is that, because Africans have been more adversely affected by Apartheid legislation than were 'coloureds' and 'indians', their current position is worse than that of these other groups. Moreover, efforts to alleviate problem areas are likely to target the worst affected problem areas first. In addition, Blacks are the largest of the groups disadvantaged by Apartheid. Black firms which engage in the construction of informal shelters (such as squatter shacks) have also been excluded from this study.

Throughout this dissertation, wherever the term 'black' or 'white' is used as an adjective to define a race group, a lower-case letter has been used. However, where the word is a noun, the uppercase letter is used. This approach does not imply disrespect on the part of the author towards any particular race group.

A further limitation concerning the sample group is that its size was limited to 100. This is largely due to the detailed nature of the questionnaire, where depth (quality of data) was favoured over breadth (quantity of data).

It should be noted here that the survey findings indicated that 'micro', rather than 'small', would have been a better description of the sample. The definition of 'small' adopted for this study was : fewer than 20 employees. However, the majority of respondents had fewer than 5 employees, while some had none.

1.4.2 Geographical Area

The survey samples were drawn only from two metropolitan areas - Durban/Pietermaritzburg and Pretori
Witwatersrand / Vereeniging (PWV). The reason for this is that the level of construction activity was
highest in these two major metropolitan areas during the research for this study. The Cape Peninsula area
was excluded because, according to personal communication between the author and representatives of th
black contractors' associations in the region, there are too few black contractors operating there to
constitute a sample of any significant size. This was confirmed by the author when efforts were made to
locate interviewees for the investigation documented in Chapter 6.

1.5 ASSUMPTIONS

It is assumed that the problems facing contractors operating in the Durban/Pietermaritzburg and PWV are
would be representative of the problems of all black SSCEs in South Africa.

1.6 METHODOLOGY

1.6.1 The Background Study

The first part of the background study (Chapter 2) addresses the ideological roots of various types of
housing delivery systems. This is considered important because SSCEs tend to be far more involved with
housing and small works than with other types of buildings. In addition, because state economic policies
affect housing delivery, types of economies are discussed, with special attention being paid to how these
relate to various underlying ideologies. The local and international literature on SSCEs is then reviewed
with the aim of establishing the current body of knowledge on: the attributes of SSCEs; the barriers faced
by SSCEs; and strategies to alleviate these problems.

The second part of the background study (Chapter 3) consists of a review the construction sector of the
South African economy and an analysis of the national and construction statistics. The main purpose of th
is to provide a contextual background for the dissertation, since its main focus is on a particular type of
enterprise operating within a specific sector of the economy. The extent to which Blacks can be identified
participants in the construction sector will be shown where possible.

The final part of the background (Chapter 4) consists of an assessment of the effectiveness of the existing
institutional support infrastructure available to SSCEs. An overview of the institutions currently supportin
SSCEs is presented in Appendix F, where details of their formation, objectives and achievements are
presented. The relevance of the inclusion of this section is that the very existence of these institutions tend
to support the hypothesis that SSCE growth is hampered by various constraints. Also, it is important to th
formulation of recommendations that the deficiencies in the existing institutional infrastructure be identifie

1.6.2   The Survey

Against this background, a survey was conducted in two metropolitan areas of South Africa to identify the attributes, background and evolution of black SSCEs, as well as the constraints that they face. The author designed and compiled the survey questionnaire and engaged research consultants to undertake the interviews. A pilot study involving five respondents was conducted in Durban. As a result, modifications were made to the questionnaire and teams were trained to conduct the interviews. The author coded the data and a research assistant was employed for the punching (which was checked by the author). The presentation and analysis of the survey data is contained in Chapter 5. In Chapter 6 the results of semi-structured interviews between the author and a limited sample of three patently successful black contractors are presented. The purpose of this is to gain a deeper understanding of the roots of success. This is followed in Chapter 7 by a conclusion in which the hypothesis is tested and recommendations are made regarding how the most serious constraints can be overcome in order that the development of SSCEs might be stimulated.

PART A

THE BACKGROUND STUDY

CHAPTER 2

A LITERATURE REVIEW

As noted in the Introduction, the thesis of this work is that the potential for black SSCEs to contribute significantly to the output of housing in the construction sector is limited by various internal deficiencies in skills, knowledge or experience, as well as by external factors and the ineffectiveness of the existing training/support/development infrastructure. This argument will be developed by (i) considering what form the economic policy changes likely to occur in South Africa's near future will take. This will be amplified by (ii) a brief discussion on types of economies. The provision of housing, while being important in its own right, is likely to be used as a catalyst for the development of the capacity of SMEs. This introduces a need (iii) to understand: 'housing' as a process; the ideologies that underpin its delivery systems; and the specific role envisaged for South African black SSCEs in housing delivery. And finally, because they are envisaged as the builders of houses, (iv) the characteristics, capabilities and barriers to growth of SSCEs will be investigated. This literature review includes international sources only where a reasonably direct applicability could be established to the local scenario. Further, the section on SSEs has deliberately been confined to the literature on construction industry SSEs.

2.1   FUTURE SOUTH AFRICAN ECONOMIC POLICY

At the time of writing, a negotiated political settlement in South Africa seems imminent. The African National Congress (ANC) will almost certainly dominate the anticipated interim government of national unity and seems equally likely to succeed in its bid to govern the country once the first post-interim government elections are held. In recent years the ANC has devoted much attention to the formulation of the policy it intends implementing. Two important objectives of this policy are: (i) to redress inequalities and injustices created by Apartheid and colonialism and; (ii) to develop a sustainable economy; with the purpose of improving the quality of life of all South Africans (ANC, 1993:1). The introduction of political democracy on the one hand and social and economic transformation on the other are clearly seen by the ANC as being strongly related (ANC, 1993:2).

ANC policy provides for compensation in respect of past inequalities and injustices through affirmative action, which is defined as : "... special measures to enable persons discriminated against on grounds of colour, gender and disability to break into fields from which they had been excluded by past discrimination." (ANC, 1993:6).

The near absence of black-owned construction MSEs and LSEs in formal markets is clearly related to the fact that Blacks were prevented by early Apartheid legislation from acquiring trade skills at artisan level. Put another way, it is argued that although this restriction later fell away, Apartheid legislation afforded Whites an unfair advantage in establishing themselves as skilled operators (and later as firm owners) in the construction indústry. This point is addressed directly in ANC policy - "Special attention will have to be

8

given to intensive training and the opening up of careers and advancement for those held back by past discrimination." (ANC, 1993:7).

The main goal of ANC economic policy is to create a balanced, growth-oriented mixed economy. In pursuance of this objective, the ANC intends implementing an economic strategy consisting of the two central components of (i) *redistribution programmes* (prioritising the provision of infrastructure, affordabl housing and basic services) and (ii) the *restructuring of the economy across all sectors* by means of grow and development-oriented strategies (ANC, 1993:12). Job creation is high on the agenda and in this rega a special role is seen for the informal sector as well as for SMEs. Although not opposed to the existence LSEs, the ANC intends introducing anti-monopoly, anti-trust and merger policies to prevent "the continu domination of the economy by a minority within the white minority.." (ANC, 1993:13).

The ANC's intention to build a mixed economy must, however, be seen for what it is - an intention. Indeed, its policy document qualifies this intention -

> "... the balance of evidence will guide the decision for or against various economic policy measures. Such flexibility means assessing the balance of evidence in restructuring the public sector to carry out national go The democratic state will therefore consider: Increasing the public sector in strategic areas through, for example, nationalisation, purchasing shareholding in companies, establishing new public corporations or joir ventures with the private sector; Reducing the public sector in certain areas in ways that will enhance efficiency, advance affirmative action and empower the historically disadvantaged, while ensuring the protection of both consumers and the rights of employment of workers." (ANC, 1993:13).

2.2   TYPES OF ECONOMIES

The above quotation suggests that the future might see a shift in ANC economic policy, depending on the success or otherwise of the initial plan. A detailed debate on the competing ideologies of socialism (and planned economies) versus capitalism (and market economies) falls beyond the scope of this dissertation. However, types of economies will be discussed since the nature of the economy can be seen to have a dir bearing on the activities of business enterprises.

Economic systems cannot be isolated from political systems, which in turn cannot be isolated from ideologies. Thus, economic systems fall within a spectrum ranging between socialistic systems at the one extreme and capitalistic systems at the other. Moll (1990:7) argues that problems have been experienced with the implementation of systems at both extremes of the spectrum - as evidenced by the failure of the 'great American dream' and the collapse of Eastern Europe and the Soviet Union - as well as in the midd - as evidenced by the collapse of the 'African socialism' adopted by Tanzania. Moll (1990:7) issues a cav that it would be a "gross error to think that the only choice available is that between capitalism on the one hand and socialism/communism on the other." While this is recognised and supported, the following section deals mainly with the ideological roots of planned (or command) economies and how, as it moves

through the spectrum (or as the elements of socialistic and capitalistic ideologies are mixed), economic
policy affects the circumstances under which housing occurs, and therefore, how it affects the firms that
build housing.

2.2.1   Planned and Mixed Economies

Beginning with socialism and planned economies, it is important to understand that Marx used the terms
'socialism' and 'communism' synonymously and that his vision was of a somewhat unreal society, when
seen from the world of today. Since Marxian socialism and what Nove (1986a:4) refers to as "really
existing socialism" differ vastly, it may be useful at this point to summarise Marx's vision.

Marx believed that capitalism retards the development of productive forces. This is evidenced by the
paradoxical co-existence, under capitalism, of surplus and unsatisfied needs. Under socialism, production
and productivity would flourish and bring an end to acquisitiveness and selfishness (Nove, 1986a:2). Marx
further contended that "commodity fetishism" and markets would fall away under socialism. Socialism
would ensure that labour power would be applied "directly...to the tasks of satisfying society's needs for
goods and services" (Nove, 1986a:3). Marx (1973:138-139) argued that it is "erroneous and absurd...to
postulate the control by united individuals of total production on the basis of exchange value, of money".
Marx also believed that the wage system and money should be eliminated. Calculations of cost would be
made in labour hours and since there would be no purchase or sale, money would be unnecessary
(Nove, 1986a:3). Marx further envisaged an end to the division of labour. "Educated, all-purpose beings
will interchange jobs. There will be no need for hierarchy, or where it is needed the task of coordination/
command can be taken in turn. The causes of alienation will be overcome." (Nove, 1986a:3). Marx saw the
state essentially as an agent of class oppression. Thus, with no class conflict, it would no longer serve any
purpose and "laws, judges, police, become unnecessary." (Nove, 1986a:4). Planning, Marx believed, would
be undertaken with the prior "knowledge of needs and of available means" and in the absence of a market,
planning would be simple (Nove, 1986a:4). Finally, Marx theorised that, under the control of the
proletariat, a classless society would emerge (Nove, 1986a:4).

Nove (1986a:4) contends that "really existing socialism" has fallen far short of Marx's vision largely but
not only because of its "utopian-romantic elements", that is: no commodity production; no division of
labour; the assumption that resources would be infinitely available; no inequality; no wages; no money; no
prices; no profit; no state; and no law (Nove, 1986b:63-64). While it is obvious that no country has yet
accomplished the (impossible?) task of instituting full-blown socialism, it may be useful to consider the case
of one that tried to - the former USSR. Deeply inspired by Marx's vision, the Bolsheviks set about
implementing socialism with its concomitant economic policies in post-revolutionary USSR. Considering the
recent reformist events in the USSR, it appears that they failed, but was this indeed the case? The following
brief résumé of the events that led to this reform will assist in an evaluation of why capitalists cite this as

an example of the non-viability of socialism and whether or not this is a fair claim.

Indeed, the socialist or planned economy of the former USSR (and the economies of its Eastern European communist allies) have failed to achieve sustainable growth, as evidenced recently by Gorbachov's economic reform process of *perestroika*. This failure occurred not because socialism is intrinsically unworkable, but because the manner in which it was implemented was *unfeasible* (Nove, 1986b:61). The sentiments of Alec Nove's encapsulate this problem concisely:

> "We should indeed be aware of the *negative* aspects of a wholly market-oriented society, and of the strength the case that can be made for planning and for social-political control democratically exercised... However, hard and realistic thinking about *feasible* socialist economies requires as a precondition, the open rejection of the utopian elements of Marx's thought. Neither he nor Lenin appreciated the inherent complexities of the problem, in its economic efficiency, social, or political dimensions. For them all would be 'simple'. We at least have no excuse for being unaware that it is all immensely complicated."(Nove, 1986b:61)

Nove's (1986b) argument is clearly that balance is important - that consideration must be given to how much socialism and how much capitalism will produce the desired results. Bukharin, the Bolshevik economist and close colleague of Lenin, saw and promoted a role for capitalism as part of a *transitional structure* from which socialism could emerge (Cohen, 1974:180), going so far as to say that, while the economic policies of 'war communism' (1918-1920) "had been militarily necessary, they were incompatible with economic development." (Cohen, 1974:140). Under 'war communism' "peasants were subject to requisitioning, a centralised authoritarian system allocated resources...and representative bodies (soviets, trade unions) were placed firmly under party control or were dissolved." (Nove, 1986a:8; Preobrazhensky 1980:21).

Bukharin saw capitalism's role in terms of what became known as the New Economic Policy (NEP). Introduced in 1921, it was 'new' because it differed fundamentally from the (old) economic policies of 'war communism'. Bukharin perceived the NEP economy as a system consisting of a public (socialist) and a private (capitalist) part. The socialist sector consisted of the banks, large-scale industry, transportation and foreign trade (the 'commanding heights'), while the private sector embraced private trade, home industries peasant farming and other pockets of private capital (Cohen, 1974:180). NEP replaced the economic policies of 'war communism', its advent being marked by the replacement of the grain requisitioning system with the imposition of a "fair tax in kind, leaving all surplus to the individual peasant" (Cohen, 1974:106). This single event led to a fundamentally different economic system and to the reinstatement of private capital, market and monetary exchange and the denationalisation of (mainly small) enterprises (Cohen, 1974:106). The state thereby became an "island in a sea of petty capitalism" (Cohen, 1974:125).

Indeed, Bukharin believed that the mixed economy of the NEP was the proper transitional structure for the emergence of socialism and, while accepting it as an interim measure, he "insisted that it was a long-term one which would serve for decades" (Cohen, 1974:180). Lenin, too, while acknowledging that NEP amounted to an acceptance that 'war communism' had failed, tried to legitimise it by pronouncing that the policies had been adopted "seriously and for a long time" (Cohen, 1974:132). In the relatively brief period before Stalin's campaign against private enterprise brought it to an end, NEP had given rise to rapid reconstruction (of the economy devastated by the war), although heavy industry lagged seriously behind the rest of the economy (Nove, 1986b:70; Cohen, 1974:136).

Stalin's rise to power brought with it the virtual elimination of private enterprise and capitalism. Agriculture was predominately collectivised and "all industry, trade, construction and transport" were nationalised (Nove, 1986a:9). In effect, 'Stalinism' replaced the NEP and introduced a

> "highly centralized economy, run on hierarchical lines, with a minimal role for markets and their spontaneous processes, with top priority for investment in heavy industry [and with] neglect of agriculture, consumer services [and] housing." (Nove, 1986a:9)

Under 'Stalinism' workers had minimal rights and free trade unions were not permitted. Peasants were subjected to "forcible collectivization" and "harsh delivery obligations" were imposed (Nove, 1986a:9). Notwithstanding his brief and wavering programme of "de-Stalinization", Stalin's successor, Khrushchev, initially supported and perpetuated Stalin's economic policies of "forcible collectivization, the excessive pace of industrialization [and] the centralized planning system" (Nove, 1986a:17). Towards the end of his rule, however, Khrushchev "encouraged Soviet reformers to develop increasingly radical criticisms of the inefficient hypercentralized system of planning and management" that characterised Stalinism (Cohen, 1986:108). Thus, despite the basic conservatism of his economic policies, Khrushchev's government represented a swing towards reformism.

The eighteen year rule of Leonid Brezhnev saw a return to Stalinism - his policies became "a wide-ranging conservative reaction to Khrushchev's reforms" (Cohen, 1986:119). Indeed, neo-Stalinism grew significantly "through the 1970s and into the 1980s" under the governments of Brezhnev's conservative successors, Yuri Andropov and Konstantin Chernenko (Cohen, 1986:139) - until the reformist Mikhail Gorbachov assumed power.

Against this background, two points of significance can be established. Firstly, at no stage in the history of the Soviet Union was socialism, as conceived by Marx, ever consummated. Secondly, the attempts of Soviet leaders to implement full socialism followed two competing courses - reformism (Bukharinism/NEP) and conservatism (Stalinism/Neo-Stalinism). Thus, it can be said that Lenin's successors were sidetracked in their quest to implement socialism by the competition between reformism and conservatism and as a result, they failed. Of these two, conservatism, with its policies of heavy industrialisation, strict central planning, and ultimate effect of economic stagnation, has left the greatest impression. The point that must be made here is that it is a mistake to equate the outcome of Soviet economic policy with pure socialist economics. However tempting it is, especially for proponents of market economies, planned economics cannot be judged on the basis of the Soviet, Eastern European, Chinese, Cuban, or any other country's experience.

2.2.2 Market Economies

The purpose of this section is to succinctly express the essence of what capitalism is and thus, how market economies operate.

In the market economy capitalism dominates - the means of production are privately owned, the wage labour system is present, and monetary exchange occurs through the market (Archer, 1988). Indeed, the *market* is the main engine of growth (Moll, 1990:8). In the market economy, the capitalist indulges in 'free enterprise', and has the legal right to pursue profit maximisation (Archer, 1988). Capital is created in the market economy through the process of production for production's sake. The social product

> "must sustain the working class and make good the depreciation of the means of production before the capitalists can gain any profits. If production did not exceed necessary consumption, the capitalist class could not exist, let alone accumulate capital, and without this class the producers would not exist as wage labourers for this category is inseparable from that of capital. Thus the regular production of a surplus is the first pre condition for the existence of capitalism." (Kay, 1975:23)

> "But...the surplus takes the form of surplus value, is appropriated by capital as profit and...ploughed back production, for it is the nature of capital to expand". (Kay, 1975:56)

The capitalist drives this process, unconcerned with its effect on society. Further, as a social system, "capitalism is not simply engaged in production; it must reproduce itself on an expanded scale" (Harrington, 1977:136). How it does this, or rather what form capital takes, is of great importance in terms of how it affects society. This aspect of capitalism and its mode of production is what Marx was concerned with. The process of accumulation is conducted in such a manner that it causes wars, depressions, *etc.* - other words, it is antisocial (Harrington, 1977:137) because, it concentrates wealth and power in fewer hands (Archer, 1988).

Archer (1988) highlights other problems with the market economy (and capitalism). It is inherently unstable, tending to 'boom' and 'bust'. It also entrenches inequality and since "much of this inequality is not required as an incentive to effort and investment...[i]t amounts...to unjust enrichment for a few at the expense of a great many others." (Lekachman and Van Loon, 1981:71). Profit can be described as unjust unearned income stemming from mere ownership of resources. However, perhaps the most crucial problem is that free markets and capitalism cannot be a global system. Capitalism extends beyond the national level to the international level - thus, in our modern world, the wealth of some nations relies entirely upon the poverty of others (Archer, 1988; Moll, 1990:37).

2.2.3 Construction in Planned and Market Economies

In this section the subjects of the construction industry in planned economies and market economies are outlined. Notwithstanding the point made at the end of section 2.2.1, the planned economies of the former Soviet Union and Eastern Europe are the best available examples to facilitate the analysis of the construction industry under socialism or capitalism. In these countries, the production of buildings was 'rationalised' (Wells, 1986:6). This involved the reduction of buildings to common components which were well suited to factory prefabrication and mass production and could be transported to sites and assembled in various ways to produce an assortment of end products (Wells, 1986:6). This practice of industrialised building depends for its success on: (i) a great degree of coordination of the demand for construction products as well as an acceptance that end products would be subjected to some degree of standardisation and; (ii) continuity of production aimed at levelling the fluctuations in the demand for construction (Wells, 1986:6-7). Thus, in planned economies, levels of construction output are tailored to the level of activity in the economy as a whole (Wells, 1986:7), or as Moll (1990:22) explains - in socialist planned economies "there is not supposed to be any phenomenon of boom or bust...[P]lanning is supposed to...guarantee consistent growth of output".

It must be noted, however, that a (capitalistic) school of thought exists which contends that industrialised building cannot produce the benefits it promises. Support for industrialised building is often "based on claims that it is faster, cheaper and better" (Strassmann, 1978:100). Strassmann (1978:100-103) argues that the claim of increased quality is ill-founded, since quality is generally improved in aspects of little importance to occupants. Further, whereas speed of *assembly* might be improved, there is little likelihood of a rise in the annual rate of, for example, dwelling production. Transport and protection of components in transit potentially become net inflators of costs, since components tend to be considerably larger. In addition, the jointing of these components becomes more complex and, therefore, more critical to the performance of the structure. Most importantly, though, sufficient volume is critical to the success of industrialised building, since deficiencies in volume negate the potential cost benefits (Van Niekerk and Hutton, 1993:567).

In the market, or capitalist economy, production in the construction sector is dependent upon demand. Since a significant part of the output of the construction sector consists of capital goods it, like other capital goods industries, is vulnerable to fluctuations in the level of activity in the economy as a whole (Wells, 1986:5). Thus, construction demand fluctuates according to the general economic cycle. In many capitalist economies, these fluctuations in demand

> "tend to be exacerbated...by the practice of governments using the construction sector as an economic regulator, cutting back on construction projects or funds for investment when the economy is overheating and deliberately stimulating investment during periods of unemployment and slack demand." (Wells, 1986:6).

This point appears to contain an inherent contradiction. By 'cutting back' in an 'overheating' economy, the government would appear to be smoothing, rather than exacerbating, the fluctuations. This can be answered by noting that, if the government pulled back *before* the economy overheated, and stimulated it *before* it reached a trough, it would smooth the cycle. However, by reacting to peaks and troughs in the cycle, the government essentially ensures that fluctuations will occur.

Typically, construction in capitalist economies is characterised by the fragmentation of the responsibility for production. Clients, design and cost management consultants, principal contractors and sub-contractors, building material and equipment producers and suppliers, financiers, *etc.* collectively produce the outputs of construction (Wells, 1986:9). This fragmentation is largely a response to the discontinuity of production and demand caused by fluctuations in the level of activity in the economy. In her studies of the construction industries in developing countries (Kenya, Tanzania and Cuba), Wells (1986:10) shows that the consequences of this fragmentation are ultimately seriously detrimental to the productivity of the construction industry. Two important consequences are: (i) that the separation of the design and construction functions makes it impossible for contractors to influence the design in terms of the appropriateness of technology and resources (Edmonds and Miles, 1984; Wells, 1986:7-10); and (ii) that moves to bring about the standardisation of products and components are hampered to the extent that this does not occur (Wells, 1986:7-10).

The fragmentation of the responsibility for construction production virtually disappears in the planned economy, due mainly to the integration of the design and construction processes, as well as the integration of the building materials and building construction industries (achieved in both cases through the prefabrication of components and standardisation of end products) (Wells, 1986:11; Strassmann, 1978:99). Thus, in planned economies there is a recognition of the fact that the construction industry and the contracting industry are indeed parts of the same industry - a recognition which is not implicit in the fragmented nature of the construction industry in capitalist economies (Wells, 1986:12). In capitalist economies, however, the focus tends to be on the contracting sector of the construction industry and a negative consequence of this is that it eliminates the potential to rationalise the organisation of resources in the building materials sector of the industry (Wells, 1986:12-13).

2.2.4 Summary

In this section (2.2) it has been noted that types of economies range between those that are rooted in the ideology of socialism and those that are founded on the ideals of capitalism. It was also noted that capitalism is a necessary forerunner of socialism. The failures of the Soviet and Eastern European economies under socialism were, however, essentially failures of economies that were only partly under socialism. *People* were dissolute - not socialism. Modern, especially Western, society equates socialism with subjugation - such has been the heritage of Stalinism - and people feel far more comfortable with some degree of freedom, particularly the freedom to accumulate capital.

Where this attitude is prevalent, it is likely that mixed economies, which have attributes of both planned and market economies, would be found. The construction industry is fundamentally affected by the degree of planning in the economy. The more market-oriented the economy becomes, the less will central planning affect the construction industry's products and the circumstances under which they are produced. Thus, in mixed economies, it is likely that the traditional procurement system would be retained, whereby design and construction are separated and standardisation is resisted.

2.3   HOUSING

It is obvious that the role of building contractors is a function of the need or demand for the built products that they are able to produce. In addition, it is equally clear that the size of a contracting firm largely determines the type and scale of building activity that it is capable of performing. In this context, smaller firms are capable of operating in the house-building sector because it is relatively (when compared with larger structures) undemanding on skills, management, working finance, *etc.* (that is, because of its relatively uncomplicated nature). By the same token, large firms are better suited than are small firms to undertake complex and high volume building work because of their advanced skills and management capabilities and access to financial and other resources. Thus, the potential for contracting firms to develop is fundamentally related to the processes at work in the delivery of the type of building products to which their capabilities are best suited. It is argued that the activities of SSCEs are essentially part of the process of the delivery of housing. This factor establishes the need to understand the arguments for and against the various theories concerning housing delivery as a necessary precursor to the formulation of recommendations regarding the development of small contractors.

2.3.1   The Ideological Debate over 'Housing'

A trend originated in the mid-1960s whereby the question of low-income housing in the third world came under new scrutiny. A major policy suggestion flowing from this was that the low-income population should be encouraged to direct its untapped labour power to house themselves (as opposed to governments assuming this responsibility) and thereby the concept of 'self-help' was promoted (Nientied and van der Linden, 1985:311). By the mid-1970s this thinking began to attract a critical response from marxists, who argued that the theoretical basis upon which self-help reasoning was based was fundamentally flawed, or at least, stood little chance of solving the housing problem. Unfortunately, relatively little real debate occurred around this conflict of ideologies. This is regrettable because certain governments and international development organisations have adopted these relatively unchallenged self-help strategies on a wide scale. For example, in the period 1980 to 1983 the World Bank allocated loans totalling US$ 1778,9 million to urban projects in 28 countries, mainly targeting "sites-and-service, squatment upgrading and integrated urban development projects" (Nientied and van der Linden, 1985:311).

To understand the potential for housing strategies to be effective, the essence of the arguments for and against self-help and alternative modes of production must first be understood. The current popularity of self-help as a housing strategy is largely due to the work of John Turner done in the 1960s and 1970s. In the words of his most notable (marxist) critic, Rod Burgess,

> "... Turner's ideas, whether he likes it or not, now form part of a growing consensus of opinion amongst housing experts, planners and international aid groups that self-help housing is not only an effective social as political palliative, but also makes good sense economically" (Burgess, 1978:1106)

Turner (1972:149-150) argues that the determination and enforcement of housing standards (what may be built and how it may be built) ultimately intensifies housing problems and results in squatting. High standards have the effect of preventing many prospective owner-builders from building their own shelter, forcing them to take out mortgage loans and to employ building contractors. Thus, the enforcement of minimum specification standards is counter-productive to the reduction of housing shortages when those who create the demand for housing cannot afford to comply therewith and also when subsidies fail to significantly improve affordability.

The concept of 'housing' as a process (verb) rather than a product (noun) is the foundation upon which Turner's (1972:151) arguments and recommendations are based. The measurement of 'housing' thus takes two forms. The measures of the *products* of housing are the specification and performance standards (that is, material or physical standards only), while the measures of the *processes* of housing are the: amount built; pecuniary costs; man-hours invested; and human effort. However, neither of these measures can quantify the degree to which needs are frustrated or satisfied. Turner (1972:152) argues that housing problems are seen in terms of material standards and housing values are determined by the physical qualit of the products and that this results in the notion that "more is better". When housing is taken to be an activity, standards become *indicators*, as opposed to *measures* of value, since only those who use houses can evaluate them (Turner, 1972:153). It follows therefore that users should be allowed to control the process, that is, they should free to choose the nature of the product, how and by whom it is built, and he it will be used.

Systems of acquiring housing constitute a spectrum ranging from controlled (closed) to uncontrolled (oper While governments of countries with low per capita incomes tend to favour closed systems, most governments implement systems which are a combination of both (Turner, 1972:155-157). Turner's (1972:158-159) view is that a positive relationship exists between a reduction of control and economy and substantiates this by citing examples of the huge savings made by owner-builders. Turner (1972:160) thus argues that the choice of system must be underpinned by the premise that housing is a process. The syste must therefore reflect an awareness of, or must include, those who participate in the process ("the actors" their actions, and their accomplishments. Three main actors can be identified - the popular (people themselves), public (government) and private (commercial) sectors (Turner, 1972:160). The actions of the participants and the products they produce are principally affected by demand, where demand is a dependent variable of their perceived needs and the means they control and are willing to contribute. The

housing process must therefore match the demand, particularly where income levels are low
(Turner, 1972:161).

Against this background, Turner's (1972) position becomes clear and can be summarised as follows. Use-values are determined by users and therefore can vary widely. For this reason it is impossible for any central government or local authority to devise a system capable of meeting all housing needs. It follows that conventional housing action fails because its products do not meet people's needs (Turner, 1972:174). The solution to this problem lies in letting the people decide -

> "My argument rests on the premise that there is no existential or real freedom without dialogue between the rule makers and the game players" (Turner, 1972:172).

And so the focus returns to the concept of self-help. Turner's (1972) position is simple - if housing delivery systems result in people remaining unhoused, then they are inappropriate. And if by helping themselves unhoused people can improve the problem, then this should be permitted and encouraged.

This kind of thinking stands accused by marxists of representing the attitudes of the petty bourgeoisie. In fact, it is not new thinking at all and its roots can be found in the industrial revolution. In the 1870s fierce debate took place between marxists and anarchists over the housing question. In particular, Engels (1979:18-39 & 75-100) analysed and severely attacked the views of Mülberger, a petty bourgeois publicist and follower of Proudhon (the French economist, sociologist and one of the founders of anarchism). In addition, Engels (1979:40-74) condemned the bourgeois views of Sax, an Austrian economist who wrote on the housing conditions of the working-class. Given the uncanny similarities between the works of Proudhon, Mülberger, Sax and Turner it may be useful to summarise the marxist reaction to these writings by looking at what Engels had to say regarding the responses of the petty bourgeoisie and the bourgeoisie to the housing question.

Engels (1979:18) differentiates between two types of housing shortages. First, there is the type that has always been around - the "bad, overcrowded and unhealthy dwellings" in which the working-class generally lives, represent a type of shortage that can be solved by abolishing "the exploitation and oppression of the working-class by the ruling class". Secondly, another type of shortage can be seen in

> "the peculiar intensification of the bad housing conditions of the workers as a result of the sudden rush of population to the big cities; colossal increase in rents, still greater congestion in the separate houses, and for some, the impossibility of finding a place to live in at all. And *this* housing shortage gets talked of so much only because it is not confined to the working-class but has affected the petty bourgeoisie as well" (Engels, 1979:18-19).

The former type of shortage results directly from the capitalist mode of production, the cornerstone of which is the creation of surplus value (Engels, 1979:19). The second type is a "*smaller*, secondary evil" which results from the capitalist mode of production, but "is not at all the direct result of the exploitation of

the worker *as* worker by the capitalist" (Engels, 1979:19). Surplus value is created when the capitalist b
worker labour power at its value and then extracts more work from these workers *than is necessary* to
reproduce the price paid for their labour. This surplus is thus produced by the working-class and taken
from it without compensation. Its distribution "among the non-working classes proceeds amid extremely
edifying squabblings and mutual swindling" (Engels, 1979:19). The petty bourgeoisie as well as all othe
classes unavoidably become a part of this process.

Engels'(1979:21) main criticism of the Proudhonist, Mülberger, is that the latter declares the housing
question to be an exclusively working-class question. This is evident from his assertion that the relations
between tenant and house-owner is the same as that between wage-worker and capitalist. Engels (1979:2
refutes this by noting that the house-owner *sells* the right to use space to the tenant and that this is
essentially a normal commodity transaction. Since the house-owner does not *employ* the tenant in order
create surplus value at his expense, his actions in this transaction cannot be described as capitalistic. Eng
(1979:22-23) argues that because Proudhon avoided dealing with and explaining the economic realities o
his subject matter, his views are irrelevant -

> "Proudhon who never bothered himself about the ... actual conditions under which any economic phenome
> occurs, is naturally ... unable to explain how the original cost price of a house is under certain circumstan
> paid back ten times over in the course of fifty years in the form of rent. Instead of examining this not at al
> difficult question economically and establishing whether it is really in contradiction with economic laws, ar
> so, how, Proudhon resorts to a bold leap from economics to jurisprudence: 'The house, once it has been b
> serves as a *perpetual legal title*' to a certain annual payment. How this comes about, *how* the house *becom*
> legal title, on this Proudhon is silent. And yet this is just what he should have explained. ...The whole
> Proudhonist teaching rests on this saving leap from economic reality into legal phraseology. Every time ou
> good Proudhon loses the economic hang of things - and this happens to him with every serious problem - h
> takes refuge in the sphere of law and appeals to eternal justice."

Mülberger (Engels, 1979:24) adopts this sense of justice and applies it to the housing question by
suggesting that it is unjust that people should be without shelter - arguing that by allowing this we make
mockery of our culture and reduce ourselves to less than savages. Engels (1979:25) rejects this as emoti
nonsense, noting that it is only due to large-scale industrialisation that the worker's bond with the land w
broken and that he was able to become "a completely propertyless proletarian, liberated from all traditio
fetters, *a free outlaw*; it is precisely this economic revolution which has created the sole conditions unde
which the exploitation of the working class in its final form, in capitalist production, can be overthrown

This brings us to the nub of the Marxist-Proudhonist dispute. While Proudhon considered the industrial
revolution to be a highly repugnant occurrence, something which should never have happened, Engels
(1979:26) saw it quite differently - rural peasants would never have been able to conceive of the idea of
huge social transformation, let alone wish to carry it out. But in their industrialised and urbanised
environment, one which for the first time ever had developed the production capacity to provide sufficie
for all and also for reserve, they could become sufficiently educated and politicised to become capable o
effecting the transformation.

Proudhonism was essentially a reaction to the conditions created by large-scale industrialisation - its solution to the housing problem was not to challenge the circumstances under which industrialisation was occurring (that is, the capitalist mode of production), but to call for the abolition of rented dwellings (Engels, 1979:29). This would save the worker from being 'exploited' by the house-owner and justice would be served. Engels (1979:29) argued that no exploitation can take place between these two parties, since no surplus value is generated as a result of the relationship. All that changes when house purchase is compared with rental is how the *already existing* surplus value is distributed.

This aspect of Proudhonism (the demand that workers own their dwellings) indeed served as a useful tool for the growth of capitalism. The marxist criticism of this revolved around the observation that property ownership had the effect of binding the worker to an employer (Engels, 1979:31), in much the same way that he was bound to the land before he was forced to migrate to the city. Thus a class of small landowners is established who then ultimately become a real impediment to "the revolutionary movement of the urban proletariat" (Engels, 1979:32).

Turning now to the big bourgeoisie and the housing question, Engels (1979:41) undertook a critique of Sax's work (which was selected because it included an attempt to summarise "the bourgeoisie literature on the subject"). Sax argued that the "propertyless classes" should be raised "to the level of the propertied classes" and that the capitalist mode of production should remain unaffected in the process (Engels, 1979:42). This thinking contains an obvious paradox -

> "... it is an unavoidable preliminary condition of the capitalist mode of production that a ... propertyless class should exist, a class which has nothing to sell but its labour power and which is therefore compelled to sell [this] to the industrial capitalists. The task of the new science of social economy invented by Herr Sax is, therefore, to find ways and means - in a state of society founded on the antagonism of capitalists, owners of raw materials, instruments of production and means of subsistence, on the one hand, and of propertyless wage-workers, who call only their labour power and nothing else their own, on the other hand - by which, inside this social order, all wage workers can be turned into capitalists without ceasing to be wage-workers... It is the essence of bourgeois socialism to want to maintain the basis of all evils of present-day society and at the same time to want to abolish the evils themselves." (Engels, 1979:42-43).

Since a shortage of housing is a necessary consequence of the capitalist mode of production, the bourgeoisie can hardly explain the existence of the shortage in terms of the mode of production. Instead, it has no alternative but to moralise "that it is the result of the wickedness of man, the result of original sin, so to speak" (Engels, 1979:44). Like Proudhon, Sax proclaims that "the housing problem can be completely solved only by transferring property in dwellings to the workers" (Engels, 1979:46). The advantages that Sax saw in this included the following (Engels, 1979:47). Firstly, the "longing inherent in man to own land" would be satisfied. Secondly, the landowner would attain the "highest conceivable stage of economic independence". Thirdly, the landowner would "become a capitalist" and because of the access to credit which the ownership of land would ensure, would be "safeguarded against the dangers of unemployment or incapacitation."

Engels (1979:47) rebutted these 'advantages', noting that "freedom of movement is the prime condition of existence" for workers in large cities and argued that no-one else, besides Sax, had discovered that city workers longed to own land. Regarding the transformation of the worker into a capitalist, Engels (1979:4 declares this to be absurd since the mere purchase of property cannot make the purchaser a capitalist, nor can it make the property capital. "Capital is the command over the unpaid labour of others" (Engels, 1979:48). Therefore, the only way in which a property can become capital is if its owner vacates it, rents to someone else and in the rental appropriates a portion of this person's labour product.

Sax, however, acknowledges that there are two type of workers' houses: the one-family-one-house-and-garden type (the "cottage system"); and the higher-density tenement type (Engels, 1979:50). Practically, these types can only exist on any significant scale, respectively, in rural and urban contexts. Engels (1979:51) identifies this as the crux of the problem with Sax's argument:

> "The housing question can be solved only when society has been sufficiently transformed for a start to be ma towards abolishing the contrast between town and country, which has been brought to its extreme point by present day capitalist society....it is not that the solution of the housing question simultaneously solves the social question, but that only by the solution of the social question, that is by the abolition of the capitalist mode of production, is the solution to the housing problem made possible. To want to solve the housing question while at the same time desiring to maintain the big modern cities is an absurdity."

Given that capital "does not *want* to solve the housing question, even if it could", the only other alternativ are self-help and state assistance (Engels, 1979:59). By Sax's own admission self-help can only accomplis anything substantial where the 'cottage system' exists already, or where it is feasible (Engels, 1979:59) - and it is not feasible in the cities, which is where the housing problem is most manifest. Thus, argues Engels (1979:64) self-help cannot eliminate the housing problem. This then leaves state assistance.

Sax identified three actions that the state could take (Engels, 1979:64-67). Firstly, it could relax legislatio affecting building standards and the free the building trades. Secondly, it could enforce hygienic condition in housing environments. Thirdly, it could use all the means at its disposal to alleviate the housing shorta (by building barracks for its own employees and granting loans for the improvement of housing condition to local authorities, societies, individuals, *etc.*). To this, Engels (1979:67-68) replied –

> "...the state as it exists today is neither able nor willing to do anything to remedy the housing calamity. The state is nothing but the organized collective power of the possessing classes, the landowners and the capitalis as against the exploited classes, the peasants and the workers. What the individual capitalists ...do not want, their state also does not want....At most [the state] will see to it that that measure of superficial palliation which has become customary is carried into execution everywhere uniformly. And we have seen that this is case."

It has been shown that: the urban housing problem; the failure to eliminate it; and the arguments for and against self-help and state assistance date back to the industrial revolution. It is not surprising therefore, th Turner's (1972) typically bourgeois views came under the same kind of attack by Burgess (1978) as did Proudhon's and Sax's by Engels (1979). The essence of this criticism has been concisely summarised by

Nientied and van der Linden (1985:313-314) as follows: (i) Turner differentiates between use- and market-values and asserts that use-values dominate. This cannot be correct because the products used to construct a self-help house (bought materials or waste materials valorised through labour) and the labour of the self-help builder assume values on the capitalist market. And since a condition of the capitalist mode of production is the "constant expansion of...commodity production", these market values actually dominate. (ii) Turner incorrectly blames the housing problem on industrialisation by not recognising that it is the capitalist mode of production that causes industrialisation. The activities of the self-help builder are part of an integral component of the capitalist mode of production - 'petty commodity production'- and capital requires this as "one of the essential conditions for the cheap reproduction of labour". (iii) Turner avoids the issue of the state's role as a ruling class and consequently "depoliticises both the housing problem and the state". He thus fails to acknowledge that the state will not intervene against the interests of capital since "it cannot reasonably be expected to legislate against the interests on which it depends and which it serves".

Turner's recommendations are therefore attractive to the state because: they have a lowering effect on wages that results from the cheap reproduction of labour; they have a quietening effect on the political demands of the poor due to the growth in social inequality; and they place the whole housing process within the capitalist mode of production (Nientied and van der Linden, 1985:314). But the state would never implement such recommendations on a significant scale because this would be tantamount to "guaranteeing access to the elements of housing" - and this, implicitly including access to land, would require redistributive policies which would "shake the foundations of capitalist society" (Nientied and van der Linden, 1985:314).

In Turner's reply to Burgess's attack, he suggests that what occurred between them was not a debate, but a discussion at cross-purposes. Turner (1978:1135) essentially argues that the passage of time and the consequent growth in experience and knowledge have broadened the parameters of the housing question debate. In Turner's (1978:1135) opinion, Engels

> 'and Marx would surely have been among the first to see the submergence of socialist potential and capitalist enterprise alike in the rise of the corporate state. The relevant debate is now between those who still see progress as the socialization of the capitalist means of production, and those who see contradictions between ways and means that alienate people from the production of basic necessities and the plentiful supply on which fulfilment depends.'

Describing himself as a conservative anarchist and Burgess as a radical authoritarian, Turner (1978:1135) considered it impossible for any real debate to occur between them unless "an honest discussion [moves] towards a symbiosis of two necessarily opposing but complementary principles: that of the 'planned society' and that of 'the good society'". Insisting that terms of reference needed to be mutually agreed upon and accepted before the debate could proceed, Turner (1978:1135) required two additional dimensions to be added to the traditional political debate between radical authoritarianism and conservative anarchism. These he identified as "the increasingly evident contradiction of corporatism and devolution" and "*scale* in space and time, or size and duration" (Turner, 1978:1135-1136).

Turner's (1978:1136) stand had much to do with his own observation that the "simplistic generalization of the principle of social ownership" results in diseconomies when taken to include housing. Consequently, Turner (1978:1136) argues for duality - central planning and local control. But his position varies according to scale and context. For example, in the British, Peruvian and North American contexts Turner (1978:1136) considers a combination of "local cooperative enterprise" and 'petty commodity' production to be both necessary and practicable. In addition to this system, a system of administration and central planning should co-exist, but should be limited to "major infrastructures and to legislation limiting concentrations of wealth and guaranteeing equitable access to locally scarce resources" (Turner, 1978:1136).

In a more direct response to Burgess and his position, Turner (1978:1135-1136) recognised the Burgess (19th century) arguments as theoretically valid for conditions prevailing at the time that they were originally conceived by Engels, but rejected them on the grounds of his own and others' 20th century experience. Turner (1978:1137) therefore prefers to adopt the concept of "*fulfilment*, through convivial modes of production on which...true economy and, therefore, social justice depend". No doubt Engels would have likened this to Proudhonism.

It seems that little will be gained by a thorough analysis of Turner's (1978) defence. What has emerged very clearly from the preceding analysis of the marxist and liberal ideologies is the following. As long as the capitalist mode of production prevails, the housing problem will exist and it is unlikely that the capitalist state will, or even can, solve it. While for marxists it is important to argue the ideology of pure socialism (as opposed to petty bourgeois and big bourgeois socialism), there is little likelihood - considering the evolution of socialism so far - that this will have any positive impact on the lot of unhoused people. If self-help housing is functional to the capitalist system and benefits the wealthy, one may well query why it is not implemented on a huge scale and why its proponents have had to work so hard to sell the site-and-service concept (Nientied and van der Linden, 1985:318). Unless the social revolution fostered by marxism occurs, and unless it occurs only once sufficient capitalism has preceded it, the victims of the housing shortage will remain the urban poor.

2.3.2   Putting Housing Strategies to Work

In the previous section it was shown that two opposing ideologies emerged in response to the housing problem. It is, however, in the field of the practical application of ideology that the solutions to homelessness will have to be found. Indeed, some authors recognise this - "...the structural critique of the self-help approach is a theoretical reaction, not an empirically grounded one" (Nientied and van der Linden 1985:318) - and warn that the gap between theory and practice should not be allowed to become too wide. Only since the end of World War II have governments across the capitalist world begun to accept that housing problems and urbanisation were related and that as institutions they would eventually have to respond in one way or another (Koenigsberger, 1986:28; Wakely, 1986:53). Koenigsberger (1986:28-31) has produced the following useful summary of the various approaches that have been adopted or promoted by the public sector between 1950 and 1985:

(a) The welfare state approach saw the establishment of "subsidised public housing...estates, new towns or extensions of existing cities...as the obvious answers". The relatively high standards of these schemes resulted in rentals exceeding the affordability of virtually all of the low-income population. By the early 1960s "practically all public housing policies, whether rural or urban, ran out of steam". Indeed, in the 1980s it has been widely accepted that large-scale government construction programmes are not a feasible option for the solution of housing problems (Slingsby, 1986:65).

(b) The design approach was an attempt to improve the image of public housing "by making up in quality for what was lacking or unattainable in quantity". It was an approach whereby designers tried to understand the needs of the low-income users of housing. As a strategy, it made little impact on the housing problem.

(c) The employer approach involved the encouragement of employers to provide both the capital as well as the rent subsidies to house their own employees. This had been tried in 19th century Britain and met with resistance because workers did not want to be tied to their employers. In the context of the 20th century and the 'third world' the opposite applied - employers resisted the approach because it tied them to their workforces. Turner and his followers interpreted the general failure of this type of approach to be due to the lack of involvement at decision-making level of those for whom the housing was being provided.

(d) The cost reduction approach emerged "after the failure of passing the buck of low-income housing to the employers". This approach brought with it the concept of incremental housing - starting with a small core and expanding it gradually at a later stage - and the reduction of standards for low-income housing. This approach failed because costs could never be cut sufficiently to make much of a difference to most of the low-income population.

(e) The mass production and prefabrication approach enjoyed some success in the United Kingdom during the reconstruction that followed the War, but this success was largely due to the significant increase in speed of this type of construction at a time when speed was important. This success was not effectively transferred to developing countries with their plentiful supplies of cheap labour.

(f) The aided self-help approach, championed by USAID, represented a joint effort between the public and popular sectors - the eventual users of the houses provided free labour and public sector agencies, who initiated projects in a paternalistic fashion, provided land, services, special skills and special materials. However, these schemes tended to be successful in rural or semi-rural settings, but not in the towns. Their contribution to the elimination of the housing problem was thus small.

(g) The site-and-service approach attracted the support of major institutions like the World Bank and differed only marginally from the aided self-help approach. This difference lay in the fact that those who were allotted plots were entirely responsible for the construction of the house. The public sector had hop that by limiting its financial involvement to 10% of the total cost of such developments, it could facilitate the provision of ten times more houses than if it had retained full control. Important as the successes wer to the concept of site-and-service, they were numerically insufficient to make a significant difference to whole housing delivery problem.

If nothing else was learnt from the approaches described above, they gave an indication of what type of action worked and what did not. However, the diversity of approaches and the tendency for new approa to have been attempts at remedying inadequacies in previous approaches, indicate that collectively these approaches constitute a failed strategy. The realisation of this has resulted in a new kind of thinking - tha responsibility for housing production should be devolved to communities. Devolution must be underpinn by the principles of (i) dweller/community *control* over the housing process and (ii) public authority acti in a *supportive*, not a controlling role (Koenigsberger, 1986:32).

Devolution has thus become the popular strategy of the 1980s and 1990s. Its acceptance as a strategy ha been gradual and has principally been due to three types of arguments. These are described by Wakely (1986:54-55) as: the 'Abrams-Koenigsberger if you can't beat them - join them argument' (whereby it is recognised that government is incapable of solving the problem); the 'World Bank argument' (whereby i makes good economic sense to use inhabitants to build their own dwellings); and the 'Turner argument' (whereby 'housing' describes a process rather than just houses and the more control users of houses hav over this process, the higher the degree of social cohesion). The 'Turner argument' was, of course, "cen to the previous two lines of argument - both intellectually and historically."

It is important to distinguish between devolution and decentralisation. The latter refers to the "decentralisation of the responsibility for the implementation of centrally taken decisions" (Wakely, 1986:56), while the former refers to "a genuine handing over and dispersal of responsibility by a central authority". Equally importantly it must be understood that devolution does not mean that government cas off or decreases its activity in housing production, it simply changes how it takes part - from controlling supporting.

Special mention must be made here of the South African government-commissioned 'De Loor Report' (SAHAC, 1992). This report on housing in South Africa proposes policy and strategies. Since it was commissioned by the government, the report takes a fairly predictable line in terms of policy proposals - private sector and foreign funds will have to pay for what the state cannot afford - which is just about everything. The report was published at a time when there was a high degree of fluidity in the political arena, and since then the negotiation process has developed to a point where it is obvious that the curren government will effectively lose control, notwithstanding its membership of a transitional executive council. There is thus little chance that the report will be instrumental in shaping future South African housing policy. Indeed it conflicts with ANC policy and has been heavily criticised in terms of: some of its basic assumptions; neglect of critical issues; and lack of insight into certain issues (see Hendler, 1993). For these reasons, the report is not referred to elsewhere in this dissertation.

2.3.3   Small-Scale Construction Enterprises and Housing

Notwithstanding the uncertainty regarding the future nature of the South African economy, SSCEs will (in terms of the ANC policy described below) play a significant role in the construction industry, which at least in the short term, will be part of a mixed (largely Soviet NEP-style) economy.

The ANC policy on SSCEs is linked to its policy on housing, of which the underlying principles are:

> "Housing is a right; ... should contribute to social equity; ... is a critical component of development;
> Community control over and participation in the housing delivery process is of the utmost importance."
> (ANC, 1993:20)

In implementing its national housing policy, the ANC anticipates diverting (current) military expenditure to housing production and prioritising investment in inner city housing and the improvement of squatter settlements, townships and rural areas over middle income housing (ANC, 1993:21; HSA, 1990b). The provision of housing will thus occur in a developmental context and will be based on sustainability in the "short-medium-to-long term" (ANC, 1993:21). Key factors in the housing strategy will be the *improvement of the capacity of small and medium-scale (presumably Black) builders* and the promotion of labour intensive construction. In this regard, it is intended that the state will carry the responsibility of facilitating the training of communities and small builders and that the private sector will be encouraged to increase its involvement in the output of lower cost housing (ANC, 1993:22). Attributes of the housing policy are reflected in ANC policy on social welfare which is founded on the tenet that "people are the fundamental resource of the country, since they have the capacity to develop personally and are central to the development of the economy and the nation as a whole" (ANC, 1993:25).

The effects of Apartheid have resulted in a situation whereby SSCEs currently face serious obstacles and will have to overcome these before they can effectively compete with the established formal contracting sector. With this in mind, it is clear that a need exists for: serious problems to be identified; deficiencies in the existing training/support infrastructure to be identified; and a strategy to facilitate the development of this group. Before proceeding with the identification of problems and the analysis of the training/support infrastructure, let us briefly outline certain points that emerged in the literature regarding the importance of SSCE involvement in the construction sector.

The ANC plans to ensure SSCEs involvement in the provision of housing. This strategy is clearly founded on various motives ranging from economic to social. Certain authors have focused on justifying this kind strategy and the main economic justifications are briefly summarised below.

(a) SSCEs are well positioned to undertake small works. In the construction sector of any country one typically finds, respectively, a small number of medium to high value (large) projects and a large number low value (small) projects at both the high and low ends of this range. The small projects tend to be spread over a relatively wide geographical area, their numbers roughly matching the population distribution. The projects tend to be undemanding, requiring a relatively low level of technological expertise, and are consequently well suited to the intrinsic capabilities of SSCEs (ILO, 1987:17). Logistically there are limi to the capability of large-scale construction enterprises (LSCEs) to undertake large numbers of small, unrelated projects. Consequently, the role of SSCEs in the construction of small projects is vital even though small projects account for a small proportion of total construction output. Equally important therefore is the need to nurture and develop these enterprises (ABC, 1990d).

(b) SSCEs could play an important role in housing developments. The widely reported withdrawal of th large private sector developers and construction companies from the South African low-income housing market (due to factors such as: high levels of violence, making construction difficult; squatters descendin on land earmarked for development; and mortgage bond boycotts) (ABC, 1989c; HSA, 1991a; HSA, 1991c; HSA, 1991d; Merrifield, 1992; UF, 1991) brought with it a vision that SSCEs could have a speci role to play in the provision of housing. While they are unlikely to be capable of performing this role unassisted, SSCEs could make a significant contribution if appropriate project management and control infrastructures were established (ABC, 1989c). It is very likely that with the advent of a new South Afric government, the return of the private sector LSCEs to the low-income housing market will be facilitated the State. The concern that this raises, is that it is far more likely that SSCEs will become involved in housing in their capacity as sub-contractors to LSCEs, than as small general contractors in their own righ As already noted, the benefits of sub-contractor involvement in housing have not been reaped - partly because of the exploitive practices of the LSCEs (Krafchik, 1990).

(c) SSEs are potentially strong generators of employment and income. SSCEs tend to utilise more labou intensive methods of construction per job than do LSCEs (ILO, 1987; ABC, 1990d) and they have the "potential to create substantial employment opportunities" (Krafchik, 1990:128). Importantly, SSCEs tend to employ more unskilled than skilled labour and the cost of creating jobs is relatively low when compare with other sectors of the economy (Krafchik, 1990:128-129). Further, they tend to be less capital-intensiv compared with LSEs. Krafchik (1990:30) has also noted that SSEs play a major role in the generation of income opportunities - "particularly to those from lower social strata".

Two important benefits could be thus be derived through the development of South African black SSCEs. These are: the greater retention of construction profits within black communities (better income distributi and; the decreases in unemployment which would result from an increase in labour intensity of SSCE construction methods. The magnitude of these benefits can be assessed if the nature of financial flows in the South African economy are studied. Black and white consumption expenditure have markedly different impacts on the economy (Roukens De Lange, 1990). Black consumption expenditure "...is concentrated mainly on the basic needs of food and semi-durable consumer goods which have low import and high employment creation effects" (Roukens De Lange, 1990). But, a significantly different pattern results from white consumption expenditure which "stimulates sectors such as transport (motor cars) and finance" (Roukens De Lange, 1990).

To summarise, it appears that if the ANC becomes the new South African government and implements its proposed housing policies, SSCE development will be considered a natural and important part of the housing process. More importantly, the State would assume the responsibility of providing or facilitating SSCE training. And if the external constraints currently faced by SSCEs can be removed, there is indeed a possibility that economic benefits would flow from this.

2.4   ATTRIBUTES OF SMALL-SCALE CONSTRUCTION ENTERPRISES

Since the focus of this dissertation is SSCEs and constraints to their development, it is necessary to understand precisely what is meant by the term SSCE. Further, it is necessary to define sub-contracting and to examine the impact of sub-contractor status on the SSCE. The purpose of this section is thus to define the broad attributes of SSCEs. A survey of the local and international literature revealed that these enterprises share common characteristics and are beset by similar problems. In this section a basis for the definition and understanding of the barriers to SSCE development will be established by drawing details from the literature of: how these firms originate; how sub-contractors originate and what their special circumstances are; what the attributes of a typical SSCE are; and what is known about the growth of SSCEs. Barriers to SSCE development are addressed separately in the next section.

Appendix G contains data collated from the journal, African Building Contractor (ABC). Over a period of three years editor Dan Padi published a regular column entitled "Success Stories", in which he reviewed the achievements of over forty black or coloured SSCEs. The author of this dissertation summarised and collated these data to create a profile of a typical black SSCE. This was successfully done. However, it was impossible to establish whether or not the data had all been collected in the same manner, or if so, what this manner was. In other words, the integrity of the data could not be authenticated. This is not to say that its integrity was in doubt, but it was, nevertheless, decided to exclude this material from the literature review. It is, however, recommended that Appendix G be read in conjunction with this section.

2.4.1   What Small-Scale Construction Enterprises Are and How They Originate

The definition adopted in this dissertation for 'small' is that specified by Krafchik (1990:4) - a firm is small when it has fewer than twenty employees. This is actually a relatively high ceiling, since it is likely that most SSCEs would retain far fewer than twenty full-time employees and instead would rely heavily on casual labour. In the construction sector, not least because of the recent proliferation of sub-contracting, many different types of SSCEs can be identified. Although collectively they are referred to as SSCEs, what makes them generically compatible is that they have a certain number of employees. Consider, for example, the radical difference between a painting firm (sub-contractor) and a general contracting firm. They could not be more different in terms of: amount of work executed per contract; level of responsibility and risk; physical and human resource requirements; *etc.*.   Yet, as long as they employ fewer than twenty people, they will both be defined as SSCEs.

SSCEs thus represent a heterogenous group. They come from a variety of backgrounds, which results in their facing different constraints. The nature and degree of sophistication of their activities vary and their business ambitions are usually considerably different. Categorisation of these enterprises is potentially problematic and international comparisons can consequently be misleading (ILO, 1987:23-24).

The problems of categorisation arise for a number of reasons. In this regard, the following should be considered. SSCEs, with the exception of the sub-contractors to LSCEs, operate at the periphery of the construction industry. At this fringe level there are frequent changes in the numbers and mix of entrepreneurs (SAHAC, 1992:178). This is usually a reflection of the relative ease of entry at the lower end of the market and of the relatively high failure rates of SSCEs (ILO, 1987:23-24). Many of these firms also operate in the informal sector, or are "occasional" contractors who enter and leave the construction sector as economic conditions and demand fluctuate. Other firms are merely sideline ventures of entrepreneurs whose principal interests are firmly rooted in sectors other than construction (ILO, 1987:24). Further, specialised tradesmen like plumbers and electricians tend to establish enterprises on an *ad hoc* basis when opportunities arise for them to undertake sufficiently large and attractive sub-contracts. Nevertheless, there are entrepreneurs who are full-time *bona fide* general contractors (ILO, 1987:24).

The ability to categorise SSCEs is clearly fundamental to the formulation of appropriate initiatives intended to facilitate their development. In the absence of any detailed South African literature on the subject, the International Labour Office's (ILO) structure for the evaluation of how SSCEs emerge in African and Asian construction industries has been adopted. It is relatively easy to enter the small-scale contracting market. technical qualifications are required and work can be undertaken with extremely low levels of initial investment (ILO, 1987:24). Further, the work at the lower end of the market is relatively technically undemanding. Consequently, entrepreneurs from a variety of backgrounds establish SSCEs. These entrepreneurs stem from two main sources, namely, the construction industry and other industrial or commercial fields (ILO, 1987:24-27).

(a) SSCEs originating from within the construction industry : Most SSCEs originate from within the
building industry. Upon analysis it is evident that entrepreneurs establish small firms via two routes - the
"trade route" and the *"management route"* (ILO, 1987:24-26).

Entrepreneurs entering SSCEs via the 'trade route' usually work their way through the ranks, for example,
a labourer goes on to qualify as an artisan, then becomes a supervisor or foreman and finally leaves his
employment to become a small general contractor. These entrepreneurs typically have a sound
understanding of practical building and are able to manage the site-based activities fairly well. However,
they tend to lack the financial and administrative skills (for example, costing, planning and scheduling,
marketing, controlling) required to run a successful small business (ILO, 1987:26; ABC, 1989c; ABC,
1989e, ABC, 1990a; Motlanthe, 1990; Merrifield, 1992). A consistent and severe weakness displayed by
these entrepreneurs is their inability to estimate and manage costs. Many make use of unreliable techniques
such as applying unscientifically compiled costs per square metre to (often inaccurate) building areas
(ABC, 1989c; Fraser, 1989; ABC, 1990a; Alexander, 1990).

Another group of entrepreneurs enter SSCEs via the 'management route', but in much smaller numbers than
those entering via the trade route. These entrepreneurs typically come from backgrounds in management
where they functioned as supervisors, clerks of works, site managers or engineers. They tend to be
successful, having had exposure to both practical site-based work as well as office-based managerial duties
(ILO, 1987:26).

(b) SSCEs originating from other industry and commerce : The group of entrepreneurs with backgrounds
in other industry and commerce who establish SSCEs can be said to enter SSCEs via the *"commercial
route"* (ILO, 1987:26-27). These entrepreneurs are attracted by the potential to the make profits, over and
above those flowing from their existing non-construction ventures, on very little basic investment. Further,
the construction market does not require them to make any long-term commitments - if their construction
ventures fail, they have the option of simply withdrawing from the sector until such time as the risks are
lower and chances of success higher. These firms seldom go bankrupt as a result of poor performance in
construction. Indeed, their main strengths lie in their ability to produce working capital (from their own
savings or cash surpluses generated by their other ventures) and their management expertise (ILO,
1987:27).

A group of entrepreneurs can be identified who have backgrounds in *building materials supply or
manufacture*. They are similar to the above group in that they have access to working capital and enjoy the
security of already being established in other profit-generating ventures. However, they enjoy obvious
advantages over the group originating from outside the construction industry - they are more familiar with
building site operations and they can minimise problems of materials supply. Consequently they are more
likely to remain in the contracting market (ILO, 1987:27).

There is, however, a tendency for these firms to confine their contracting operations to the particular geographical area of their materials supply or manufacturing businesses. This usually reflects a desire to maintain reasonably tight control over their other interests and a reluctance to delegate control of the contracting operations to employees. For obvious reasons, this geographical area tends to coincide with the radius over which they can supply their own building materials economically (ILO, 1987:27).

(c) SSCEs originating from the ranks of the unemployed or retrenched: The findings of Krafchik's (1990:87) case study revealed that over half of his respondents had been unemployed prior to starting the sub-contracting firms. This is evidence of the relative ease of entry into the construction sector. It also suggests that many black SSCEs might be 'pushed' into self-employment involuntarily - they are retrenched and have little option but to set themselves up as sub-contracting firms specialising in the various trades.

2.4.2   Sub-contracting

Much interest has been shown in the sub-contracting phenomenon. The issues that constitute the subject are many and varied, but there does appear to be a connection between sub-contracting and formal employment. Krafchik (1990:87) has defined this as follows. There are three categories of reasons why sub-contractors started their firms - negative reasons (pushed into the sector), positive reasons (pulled into the sector) and neutral reasons (inherited firm). By far the largest of these categories, accounting for over half of all reasons, consists of individuals that were "pushed" into sub-contracting through a "lack of employment alternatives". Krafchik (1990:87) further noted that "a common mechanism in large firms was to offer workers a subcontract as an alternative to retrenchment". Sub-contracting enterprises thus frequently originate as a result of large contractors having adopted a policy of overhead-minimisation (ILO, 1987:63; Krafchik, 1990)

Given the connection between sub-contractor status and unemployment/retrenchment, it is necessary to briefly discuss sub-contracting and the issues associated therewith. The definition of what a sub-contractor is can be formulated by considering the nature of a general contractor. General contractors usually contract directly with the party commissioning the work and they assume responsibility for the construction of a complete job. The manner in which the work is executed is thus largely the firm's decision. The firm may elect to employ smaller specialist firms to execute portions of the work or might employ staff with the necessary skills and perform the work itself. In these cases, where general contractors employ other firms to do their work, they are referred to as 'main contractors' or 'principal contractors'.

Sub-contractors are employed by principal contractors. However, the basis of the sub-contracting relationship is negotiable and may take two main forms, namely, labour-and-material and labour-only sub-contracts.

A *labour-and-material sub-contractor* is usually employed by a general contractor to perform a specialised element of the building work (for example, the electrical installation, plumbing and drainage, roof tiling, *etc.*). The sub-contracting firm supplies its own labour and materials, but by agreement would make use of some of the general contractor's plant (equipment). Another typical type of labour-and-material sub-contractor is the general contractor who is employed by a developer to build complete buildings, with the developer effectively acting as project manager.

A *labour-only sub-contractor* is similar to the labour-and-material sub-contractor, but the firm supplies labour only. Further, the nature of the work it performs is usually not particularly specialised (for example, painting, plastering, rough carpentry, bricklaying, *etc.*). These sub-contracting firms are usually owned by 'skilled' artisans, often former employees of large general contractors.

It was noted above that many firms resort to, rather than choose, sub-contracting. It is therefore only appropriate to categorise an enterprise as a sub-contracting firm if, by its own design, it undertakes sub-contracts predominately. This is usually the case with specialist firms, chiefly electrical and plumbing contractors, which would never seek to take on full contracts as general contractors. However, it must be recognised that the most feasible way of identifying sub-contractors is to define them as those firms which are in a sub-contracting relationship at the time of observation.

An interesting point emerged from the articles discussed in Appendix G. In general, "bricklaying" and "building" were perceived by South African SSCEs to be the same activity. Thus, many individuals describe themselves as "builders" once they have acquired the skill of laying bricks. It consequently becomes difficult to differentiate between bricklaying sub-contractors and small general contractors.

Sub-contracting has become an established part of the South African construction industry. Over the period 1972 to 1985, gross output in the construction industry produced by sub-contractors increased from 35% to 45% (Krafchik, 1990:72). These proportions are probably understated, given that the majority (almost 70%) of Krafchik's (1990:76) sample of sub-contractors were not registered with Industrial Councils. Another reason why these proportions are probably understated is that the output of an unregistered sub-contracting firm is recorded as part of the output of the registered principal contractor (Krafchik, 1990:71). In addition, both Merrifield (1992:66) and Krafchik (1990:76) reported from their interviews with contractors and sub-contractors that the prevalence of sub-contracting is widely acknowledged in the South African building industry, especially in the black housing industry.

Whether or not sub-contracting firms are employed by LSCEs depends on the numbers of firms available as well as their competence and reliability (ILO, 1987:64). ILO research shows that the practice of sub-contracting is generally not widespread in developing African countries (ILO, 1987:64), while in Asia it is common practice, with labour-only sub-contractors (called "mandors") being a traditional feature of the Indonesian construction industry (ILO, 1987:25). However, it was stressed that each national or provincial setting needs to be studied separately to determine "the numbers, specialities, competence and capacities of small-scale sub-contractors" and the potential for the promotion of sub-contracting (ILO, 1987:65).

Sub-contractors who do not choose to operate in that capacity generally desire to become independent of principal contractors, citing lack of access to land as the reason why they remain sub-contractors (Padi, 1988a; Padi, 1990b). Further, Merrifield (1992:61) suggests that most sub-contractors express "a strong desire to become general contractors" because they wish to escape the exploitive tendencies (for example, artificially low prices fixed by developers (Motlanthe, 1990)) of principal contractors. However, there is evidence that some sub-contractors recognise their limitations and, indeed, prefer sub-contracting (Padi, 1990c).

It is not the purpose of this dissertation to debate the economic, social and political aspects of sub-contracting (Krafchik (1990) has covered most of that territory). Rather, sub-contracting is identified as an established feature of the South African construction industry and, as such, the circumstances under which it occurs, or could occur, are of concern. The primary argument for the promotion of sub-contracting is that it results in a greater number of small firms obtaining access to work opportunities than would otherwise have been the case (ILO, 1987:63; Krafchik, 1990:128). Black SSEs in South Africa are virtually excluded from the mainstream of the economy and "access lanes" need to be identified, cleared, or created (Rolfe, 1990). Sub-contracting is certainly one of the more obvious ways whereby Blacks could achieve greater participation in the construction industry.

It has been shown that sub-contractors generally perceive themselves to be the victims of, or at least vulnerable to, exploitation by principal contractors. However, it is likely that in many cases, principal contractors penalise sub-contractors over issues concerning quality of work. Indeed, the benefits to principal contractors of issuing sub-contracts depend heavily on the sub-contracting firms being competent and reliable (ILO, 1987:64). South African employers of construction industry sub-contractors suggest that black (particularly labour-only) sub-contractors leave much to be desired when issues such as professionalism, quality of work, productivity and ability to work without supervision are considered (ABC, 1989a; ABC, 1989e; ABC, 1989c; Motlanthe, 1990; Merrifield, 1992).

Against this background, it appears that the South African construction industry would benefit by initiatives aimed at strengthening the position of sub-contractors (Merrifield, 1992). Such programmes must, however, be carefully designed, giving due regard to the terms of the sub-contracting relationship. Krafchik (1990:180) warns that, *in its present form* in the Western Cape:

> "The sub-contracting relationship serves to inhibit employment and income generation in the sub-contracting firms in three ways. These are rates of pay, shortages of raw materials and insecurity of tenure. The cause of these constraints is largely the degree of inequality inherent in the sub-contracting relationship... ."

A possible solution might be found in the anticipated changes in State land allocation policies that might enable contractors other than the largest developers to obtain land. This would result in sub-contractors being employed by medium-scale contractors, as well as by individuals. Sub-contractors would probably retain greater profits and their bargaining power with big developers would be strengthened (Krafchik, 1990:180).

2.4.3   Typical Attributes of Small-Scale Construction Enterprises

Box 1 presents a generic profile of SSCEs. The outline largely represents the profile compiled by participants in an ILO training course held at Kuala Lumpur in 1982. There were clear similarities between this and the South African situation. The outline has therefore been adopted and expanded to include typical characteristics of South African black SSCEs.

It is common for SSCEs to fail at an early stage. Either they are forced into bankruptcy, or they withdraw voluntarily in the face of declining profitability (ILO, 1987; ABC, 1990a). These enterprises are particularly vulnerable to downturns in demand, which sometimes precipitates the failure of once successful and long established firms. As demand drops, firms tend to experience long periods of inactivity. Competition increases, putting downward pressure on mark-ups. Notably, failure of SSCEs results as easily from a series of minor errors of judgement on successive jobs as it does from big losses on one or two jobs (ILO, 1987; Padi, 1988c; Alexander, 1990).

2.4.4   Growth and the Small-Scale Construction Enterprise

There are some SSCEs which either choose, or are forced, to remain small. Indeed, many of these contractors have neither the ambition, nor the ability to expand. In these cases the firm expands only until it produces an acceptable personal income for the owner (ILO, 1987:28). The decision to remain small often reflects the owner's desire to minimise administration or to continue doing the skilled work personally. It frequently, however, reflects the owner's fear of losing control and his consequent reluctance to delegate to employees or partners (ILO, 1987:28; Fraser, 1989; Ferreira and Potgieter, 1989).

Certain SSCEs, particularly in populous urban areas, remain small because they can survive by operating in fairly small catchment zones. This is usually possible because these areas generate a demand for new buildings as well as for the maintenance and alteration of existing buildings (ILO, 1987:28). In fact, in recent years the alterations and additions market has been a very important contributor to gross domestic fixed investment (GDFI) in buildings (HSA, 1989b). Indeed the additions and alterations market in the Western Cape was the "only segment to exhibit real growth" during the 1980s. And it can be expected that

The SSCE owner leaves school at approximately sixteen years old without having completed his secondary schooling. Immediately thereafter he enters the construction industry as a general worker or he joins the family business as an apprentice in his father's trade.

While in this employ, the SSCE owner advances to the level of skilled artisan, often without formal qualification, and gains some supervisory experience (this involves the supervision and control of his own team of assistants or he rises to the level of foreman).

The entrepreneur then becomes disenchanted with employee status or concludes that his prospects of entering into partnership with his employer are remote and leaves to start his own business, operating from home. Initial sources of finance are his own savings or family assets. (In the majority of cases this is used as working capital for his first contract. Less frequently these savings provide sufficient collateral for a bank loan).

The SSCE owner employs a small full-time staff, seldom exceeding a total of 9 unskilled and skilled members, but also employs casual labour on an *ad hoc* basis.

Initially the SSCE's clients are happy with the firm's performance. As his work load increases, the SSCE owner begins to experience difficulty in managing his business. This results in a drop in quality and the firm's reputation deteriorates. The SSCE owner has little knowledge of accounting and costing and this compounds his management problems

There are always some clients who are slow in paying and the SSCE experiences cash flow problems, often eventually causing the business to fail. Cash flow problems are aggravated by delays in the supply of materials.

Sources: *(ILO, 1987; Fraser, 1989; ABC, 1989a; ABC, 1990a; ABC, 1990c; Motlanthe, 1990)*

Box 1   Profile of a typical SSCE

this growth will continue (Krafchik, 1991:3).

Many SSCEs do, however, express a desire to grow. However, there are important differences between construction enterprises and non-construction enterprises. In the latter, expansion would usually be preced by market research, additional investment in premises and means of production, *etc.*. Or, additional mea of production could be acquired via company take-overs and mergers. These tactics are not commonly employed by expanding construction SMEs. Because of the *in-situ* nature of the production process, expansion is unlikely to require new investment in factory premises. Further, if expansion necessitates the use of heavier equipment, it can usually be hired and paid for out as part of current expenditure. Take-overs among SMEs are infrequent (ILO, 1987:29-30).

SSCEs grow via the avenues of *'lateral expansion'* or *'upward movement'* (ILO, 1987:30). The former p refers to the firm doing more work, while the latter refers to it doing bigger, more complex jobs. However, some firms adopt both of these approaches simultaneously, but this happens relatively infrequently. With lateral expansion comes increased stress and pressure on management, particularly for the firms run by just one person. The obvious way of reducing this stress would be to take on partners. But, reportedly, suitable individuals with construction experience are difficult to find. Lateral expansion initiatives are risky and could easily result in bankruptcy (ILO, 1987:30; Merrifield, 1992). Growth through upward movement is also potentially problematic, with management coming under pressure for different reasons. The level of technology required on the larger contracts is often much more complex, falling outside of the firm's previous experience. As a result, many SSCEs find that they are unable to cope. To compound matters, these larger contracts require larger labour forces and a greater intensity of output. The smooth coordination of the construction activity becomes critical, involving a variety of elements of the building, sub-contractors and specialist installations (ILO, 1987:32). Those contractors who overcome these difficulties frequently fail to produce acceptable quality and the cost of remedial work can quickly consume profits.

It can be seen that the realisation of growth presents a considerable challenge to the management of the SSCE. For growth to occur, it is principally the quality of management that must improve and this places considerable demands and strain on managers (Merrifield, 1992). In addition, most growth attempts are frustrated by a lack of access to capital (ILO, 1987; ABC, 1989e; Motlanthe, 1990).

2.5 FACTORS CONSTRAINING THE GROWTH OF SMALL-SCALE CONSTRUCTION ENTERPRISES

The constraints to SSCE growth are categorised in this section in three groups. First, there are internal factors, or management constraints. Secondly, there are the external factors, or market/client-related constraints. Finally, there are more general external constraints, racial issues, *etc.*. The emphasis in this section is once again on the consolidated experience of the ILO in Africa and Asia, but where appropriate, similarities deriving from the local situation have been added to expand this.

2.5.1 Internal Factors

In this section internal constraints are taken to be those that the SSCE owner can or could have remedied, that is, these are inadequacies of the contractor himself. The most apparent of these are financial management and education and training (Krafchik, 1990). But, clear areas of inadequacy exist within these two broad categories - hence the specificity of the data below.

(a) Inability to understand drawings and specifications : Most SSCEs experience problems in

understanding technical drawings and specifications (ABC, 1989b; ABC, 1989c; Fraser, 1989; ILO, 198
However, this improves over time, but the ability they acquire seldom extends beyond the reasonably
narrow range of their experience (ILO, 1987:45).

(b) Inability to estimate costs, compile tenders and assess the effects of inflation : This is a universal
problem for SSCEs and clearly reflects their lack of training or experience in business and financial
management. In the absence of this expertise, SSCEs tend to rely on intuition (based on their previous
limited experience), commonly underestimating labour productivity and material transport costs as these
vary from one contract to another (ILO, 1987:45-46).

This problem appears to be widespread among black SSCEs. Many regard their inability to compute and
manage costs as their single most critical inadequacy (ABC, 1989c; Fraser, 1989; ABC, 1990a; Alexand
1990; Creighton 1992; Merrifield, 1992). The following comments by Fraser (1989) give an impression
the situation,

> "The lack of costing skills has led to underpricing of contracts and misinterpretations of terms of the contra
> An African builder also faces heavy financial losses at the end of the project by virtue of the fact that he fa
> to incorporate costs associated with overheads or contingencies in compiling and quoting for tenders.
>
> What most African contractors do, and are very confident of, is the use of a standard rate per m² as a mea
> of estimating. This is further reinforced by the popular consensus as to what constitutes an acceptable town
> rate and the willingness of competitors to undercut any contractor who tries to increase his rates.
>
> This method of pricing leads most contractors to end up underpricing, since they tend to use the same rate
> all their projects, irrespective of the type of finishes, structure, allocation and nature of foundations. To
> mention the worst part, these "township" rates, in some cases, have remained unchanged for the past five
> years, irrespective of the inflation prevailing today and the real value of a rand in the economy."

The apparent lack of understanding of inflation and building material price escalation clearly presents an
imposing barrier to black SSCEs wishing to compete in the more formal house building market. Building
societies in South Africa are reluctant to allow their home buying clients to employ builders who insert
escalation clauses in their contracts, since they wish to protect these clients from increases in originally
agreed costs. However, this essentially forces the contractor to estimate price increases in advance and to
include an amount for this in his tender (Griffin, 1990). In other words, the contractor would have to ass
and cover the *risk* of price increases. Merrifield's (1992) has noted that black SSCEs are generally unabl
to manage business risks.

(c) Unfamiliarity with legal aspects of contract work : SSCEs lack the sophistication of the larger
contractors. This is particularly evident in their inability to use the conditions of contract to their advanta
Clients almost invariably alter the basis of original contracts via the issue of variation orders and
instructions. A thorough knowledge and understanding of the contract is vital to the successful negotiatio
of rates for variations, where considerable financial gains stand to be made. It is, however, frequently th
case that SSCEs lose money due to poor preparation and negotiation of claims against contract variations

(ILO, 1987:46-47).

(d) Lack of financial record-keeping : SSCEs generally tend to neglect bookkeeping. Many see building as the firm's main function and consequently downplay the administrative aspects of running the business. Few of these contractors seem to recognise the positive role that accurate cash flow records, *etc.* could play in attempts to raise working capital from institutions (ILO, 1987:48-49). Krafchik's (1990:228) study of Western Cape sub-contractors suggests that this is also true of black SSCEs where he found that 59% had not kept financial or other such records at all.

(e) Inability to plan and control projects : The ILO (1987:47) regards "deficiencies in planning and management skills [as being probably] the greatest single stumbling block among small-scale contractors". Common examples of these deficiencies are the inability to: "compile materials procurement schedules; institute productivity checks during contracts; anticipate possible delays; plan transport requirements" and so on (ILO, 1987:47). Poor planning and project control have also been raised as issues of concern in the South African context (ABC, 1989e).

SSCEs clearly do not need the sophisticated management tools that larger contractors commonly use. Simple bar charts or activity schedules would suffice in most cases. However, the use of planning techniques is critical at SSCE level, since profits are usually small and consequently, margins for error are slender (ILO, 1987:48).

2.5.2   External Factors

The external factors constraining SSCE growth flow from legislation as well as the business environment, markets, clients, *etc.*.

(a) Restrictive legislation : It must be noted that legislation affecting black SSCEs falls into two categories - repealed and current. But within each of these categories there is also a division - between legislation affecting Whites and Non-Whites. It is not the purpose of this section to present a detailed review of past or current racial legislation (for example, influx control, group areas), but the laws of the past have caused enduring damage. The more important of those that have affected black SSCEs are mentioned below.

Many of the bricklaying contractors interviewed by Padi (see Appendix G) had been harassed by officials for 'illegally' handling trowels in the 1950s (ABC, 1988c; Padi, 1989a; Padi, 1989b; Padi, 1990b; Padi, 1990c). Clearly, black artisans have historically been relatively less exposed to positions of responsibility and the practice of skilled work than have their white counterparts. Thus, it could be argued that there would currently exist larger numbers of competent black SSCEs if their owners had they inherited the skills and accomplishments which were denied their forebears. Merrifield (1992:60-61) supports this argument, noting that:

> "[b]lack builders and sub-contractors are labouring under many limitations, most of which are a consequence
> of Apartheid, and are thus finding it difficult to establish themselves in the market. ...As a result of Apartheid
> black builders were restricted from the building trades, and, while these restrictions began to be lifted in the
> 1970s, they still find themselves struggling to be accepted as qualified tradesmen."

Perhaps the most important historical barrier to black SSCEs seeking entry to the formal building
environment was the provisions of the BIFSA by-laws. BIFSA's members are "any natural or juristic
person who is a member of an affiliated association" (BIFSA Official Handbook) and the majority of BIFSA
affiliates are the members of the Master Builders' and Allied Trades Associations (MBAs). Until 1988,
BIFSA members were bound in terms of By-Law 5.1.1 to refuse to tender in competition against non-
BIFSA members. The Financial Mail observed that "this provision ha[d] virtually forced developers over
the years to exclude all non-MBA members from tenders since most of the large builders [were] MBA
members" (Financial Mail, 1986). It is only since January 1988 that BIFSA By-Law 7.3 has stipulated that
members may not embargo clients who call for tenders from non-members in competition with its members
(BIFSA, 1988b). This change, however, had a lot more to do with the investigation and ruling of the
Competitions Board than it did with BIFSA's benevolence. In fact, BIFSA and the South African Property
Owners Association (SAPOA) had agreed between themselves to *retain* the cartel-style provisions of By-
Law 5.1.1 in their proposal to the Competitions Board (Financial Mail, 1986).

We now turn to the implications for SSCEs of the Labour Relations Act (1956) (see Appendix B for more
detail). The vast majority of black SSCEs are not registered with Industrial Councils (Krafchik, 1990), but
they can only operate in this capacity as long as they can elude detection by Industrial Council inspectors.
And detection of non-registration is less likely in informal markets. The choice to stay unregistered is thus
essentially a choice to stay out of the formal markets, where detection of non-registration would certainly
guaranteed.

It is important to note that all employers in the building industry are bound to comply with (virtually all of
the conditions of the Industrial Council Agreement operating in the area - whether or not they are members
of the employer or employee bodies that were party to the agreement. Krafchik (1990:83) questioned sub-
contractors on why they had not registered with the Industrial Council and reported that: they considered
the costs of registration to be too high; they did not see any benefit in registering; and they found it easier
to obviate the regulations governing prohibited employment practices and minimum wages. Compliance
with the agreements is thus seen as expensive and/or restrictive.

One cannot begin to quantify the effects of the Group Areas Act and Influx Control legislation upon Blacks
in general, or black SSCEs in particular. At the very least, it should be clear that it is no accident that black
SSCEs have tended to build in 'black areas' and white SSCEs in 'white areas'. Even now that the
legislation has been repealed, its effects will continue to be felt. 'Black areas' are not going to 'turn white'

overnight, nor *vice versa*. And it is from these areas that the demand for the services of SSCEs flows, that is, the existing communication networks through which work is acquired still reflects the situation that existed prior to the repeal of the Group Areas Act.

(b) LSCE monopoly over the housing market : The larger developers, who have the capacity to construct hundreds or thousands of houses simultaneously, have established themselves in the housing market. But, this is initially the easiest market for black SSCEs to enter. They are thus in competition, but are fundamentally unequal. In most cases these developers are white and are better organised and resourced than are black SSCEs - who will have to prove themselves and aggressively market their capabilities if they are to become established in this area (Padi, 1988a; ABC, 1990b; Padi, 1990b; Creighton, 1992).

Not only are the LSCEs better organised and resourced than the SSCEs, but they also tend to deal with relatively more sophisticated clients. SSCEs often have to contend with problems originating from clients. For example, drawings and specifications are often incomplete or deficient in detail because clients have assumed that contractors know what to do. This often results in important costs being omitted in the SSCE's tender (ILO, 1987:42). An interesting study was conducted by a South African estimating bureau involving the analysis of 400 drawings that had been submitted to it for estimating on both new houses and alterations contracts in the Black housing market. This study revealed that:

> "approximately 20% of the drawings had no specification at all; approximately 60% ... had only a minimum standard specification - usually pre-printed on the plan; a very small percentage of plans ... included an acceptable specification and were usually from the larger housing companies and government departments; approximately 54% ... were inaccurately drawn so that measuring was made very difficult or even impossible; and approximately 15% of plans presented were not possible or practical to build - 90% of these cases applied to roof structures" (Alexander, 1990).

In addition, inexperienced clients tend to be pedantic about accuracy and quality of workmanship. This often results in SSCEs being forced to perform very costly remedial work when less expensive compromises could possibly have been reached. This tendency presents a serious constraint to the development of domestic contracting industries (ILO, 1987:42).

It can thus be seen that the LSCEs have the capacity and the track record to obtain the bulk of the housing development work. It can also be seen that black SSCEs are hampered, by the lack of sophistication of the clients they tend to deal with, in their attempts to build up track records. And this ensures that the established firms retain their advantage.

There is an acute shortage of land available for development in South Africa. It must be stressed though, that this is more a problem of supply than availability and is largely due to Government land allocation policy (HSA, 1991a; Krafchik, 1990:160). Black contractors seeking sites to develop seldom meet the financial requirements demanded in the allocation of large numbers of sites (Creighton, 1992). Consequently, they either become sub-contractors, or they acquire a small number of sites from a local authority, contractors' association or individuals. A useful insight into this problem is provided by an instructional article in African Building Contractor, intended for a readership of black SSCEs:

> "Land can be made available to a builder in various ways : (i) From a large company which is the employ[er] the individual who wishes to build, (ii) a township developer, (iii) a local authority such as [a] town counc[il] (iv) an independent landowner" (ABC, 1988a).

Implicitly the article tends to entrench the almost unchallenged relegation of black contractors to employ[ee] or sub-contractor status. In black contracting circles it is, indeed, accepted as fact that only large white developers will get access to land for the development of black townships (ABC, 1990b).

(c) Irregularity in the inflow of work : SSCEs tend to experience perpetual difficulty in obtaining work [on] a steady basis. Indeed, so do all other contractors, but SSCEs are particularly vulnerable in times of low demand. Further, the absence of national policies on the allocation of work for small firms tends to ensu[re] that this remains the case. Public contracts are usually issued on a pre-qualification system which categorises and restricts contractors to maximum contract values they are permitted to undertake (ILO, 1987:37).

(d) Lack of access to finance : Access to finance is a major problem for SSCEs, chiefly due to their inability to meet collateral requirements (ILO, 1987:38). South African black SSCEs are generally rega[rded] by financiers as falling into a high risk category (Mngomezulu, 1988; Padi, 1988a; ABC, 1989e; Frase[r,] 1989; Padi, 1989c; Alexander, 1990; Padi, 1990c; ABC, 1991b). Reasons why this is the case include: absence of a good track record (Motlanthe, 1990); lack of sophistication in the preparation of applicatio[ns] and non-membership of contractors' associations (Alexander, 1990). SSCEs generally, and black SSCE[s] specifically, find it almost impossible to obtain credit from building material merchants - also largely d[ue] the absence of track records (ILO, 1987:40; Alexander, 1990; Merrifield, 1992). A further complicatio[n] the South African scenario is the state of political unrest. Financial institutions are loathe to accept risk[s] which they are unable to assess and over which they have no control. There is, thus, a tendency for the[se] institutions to avoid financing black housing (Nonyane, 1988; HSA, 1991b). It follows that black contractors similarly experience difficulty obtaining loans for their involvement on such projects.

The consequences of limited access to finance tend to have a detrimental effect on the reputations of SS[CEs] who often resort to taking deposits for new contracts as a means of financing their current work (Alexander, 1990; Motlanthe, 1990). This tendency is partly a reaction to the contractual conditions governing payment, which often present great difficulties for SSCEs. It is common for there to be fairl[y] lengthy delays between the time of certification of work completed and the receipt of cash, particularly [in] the case of public sector clients (ILO, 1987:43). In South Africa, most of the problems regarding paym[ent] derive from the principal contractor:sub-contractor relationship. In addition to the complaint that initial 'tender prices' are often fixed by the principal contractor (Motlanthe, 1990), many black SSCEs compla[in] of problems in their subsequent financial relationship with principal contractors (Padi, 1988a; Padi, 198[9]

Examples of these problems given in Krafchik's (1990:166) study included: undue delays in payment; retention monies not being refunded; rates for contracts being reduced without the prior agreement of the sub-contractor; and excessive penalties being applied - often for absurd reasons.

(e) Shortages of skilled labour : The shortage of skilled (Appendix B defines 'skilled' and who may legally perform skilled work) labour on building sites is a problem facing SSCEs the world over (ILO, 1987:40). The input of skilled workers and foremen plays an important role in whether or not SSCEs produce quality work, and therefore, whether or not they operate profitably. It is possible that the apparent shortage of skilled workers at small firm level is a reflection of the reluctance of such workers to work for SSCEs. Reasons for this may be that jobs are small, pay is poor and security of employment is not guaranteed (ILO, 1987:40).

Shortages of skilled workers also affect South African contractors. Whether or not this is a problem, must be seen from the contractor's outlook. For example, of the forty-three black SSCEs interviewed by Padi (see Appendix G), only one specifically mentioned that "skilled workers [were] a problem and very unreliable" (Padi, 1990b). This conceivably indicates that the majority of Padi's black SSCEs were actually using unskilled workers to perform skilled work. This hypothesis tends to be confirmed by Merrifield's (1992:68) finding that "most skilled or semi-skilled tasks on site are performed by operatives with only on-the-job training". Similarly, Krafchik (1990:126) found in his survey that 77% of those actually performing skilled work were not qualified artisans.

(f) Lack of access to economies of scale and materials supply problems : As noted above black SSCEs lack the capacity to compete with the larger white developers and contractors. Merrifield (1992) has noted that a turnover of approximately 25 houses per month is required before it becomes viable for white contractors to operate in the low-income housing market. But Merrifield (1992) has further noted that black contractors working under the various development agencies and programmes (see Appendix F) are often restricted to a turnover of 6 to 18 houses per year. Merrifield (1992) describes the consequent denial of access to economies of scale as the most critical problem facing black contractors.

Indeed, the SA Housing Trust (SAHT), argues that its motive for initially having employed established LSCEs on its projects was that, due to their higher turnover, they had access to bulk buying discounts which were unavailable to SSCEs (ABC, 1989a). This is a somewhat specious argument, since it would clearly have been possible for the SAHT to have purchased materials on behalf of SSCEs. Nevertheless, the issue of lack of access to economies of scale is reported to be a problem experienced by all SSCEs (ILO, 1987:40).

The problem of the erratic and unreliable supply of materials to sites is also commonly experienced by SSCEs. Often this is the result of the poor planning of material requirements. However, some black labour-only SSCEs report that the principal contractor, who provides the materials, is responsible for delays in the

supply of materials. This suggests that the materials merchants are to blame and therefore, that the eff
planning of material requirements would not necessarily diminish the problem of inefficient supply (IL
1987:40; ABC, 1989a; ABC, 1990d; Krafchik, 1990:163; Merrifield, 1992). Further complications st
from the inability of most SSCEs to obtain credit from merchants. This has the effect of placing huge
demands on the need for working capital which, as noted above, cannot easily be obtained through for
sources.

(g) Lack of access to hired plant and equipment : SSCEs do not usually encounter the need for the la
items of plant (for example, concrete mixers) until they start undertaking large contracts. Initially, mo
capable of operating with just their basic hand tools, wheelbarrows and perhaps a pick-up. Many still
concrete and mortar by hand (ILO, 1987:40). This tendency to adopt manual, rather than mechanical,
methods of building might well be due to the SSCE's inability to assess whether or not productivity w
improve sufficiently to justify the expense. And hire charges do tend to be expensive, which may well
deter SSCEs from obtaining plant.

(h) Lack of access to and quality of training : The majority of employees in the South African buildi
industry work for employers who are not members of BIFSA. Consequently, they do not have ready a
to the BIFSA skills training facilities and most training therefore occurs on the job (Merrifield, 1992:6
this context, the following results, from a BMI national survey (Merrifield, 1992:69), are of interest. I
a sample of 141 South African home builders, it was found that 43% believed that the training of artis
should be the responsibility of the individual himself. Further: 26% felt that training was the governm
responsibility; 24% felt that principal contractors and developers should be responsible for training lab
on their jobs; 13% thought that sub-contractors should train their own employees; and only 11% saw
training as BIFSA's responsibility. Merrifield (1992:69) concluded that LSCEs are essentially unintere
in how or where their employees obtain training.

Management training is virtually unavailable to most SSCEs and those programs which do exist (see
Appendix F) cannot cope with the demand for this type of training. In addition, entry levels are often
prohibitively high for many black SSCEs, particularly because of their generally poor track records. I
addition, SSCEs operate in a market characterised by slim margins, and generally cannot afford large
training budgets (ABC, 1990c). These training issues will be discussed in much greater detail in Chap

2.5.3 General Factors

(a) The poor reputation of contractors : Contractors, especially in developing countries are frequentl
criticised for: their tendency to produce "poor quality work"; their "failure to meet completion deadlin
and their "unreliability in undertaking remedial and completion work" (ILO, 1987:35). This is, indeed

of black SSCEs and is a matter of great concern to their representatives and leaders (ABC, 1989a; ABC, 1989e). In South Africa there is, however, an additional factor. Not only do *small* contractors have a poor reputation, but *black* small contractors have a worse reputation than other small contractors. The issue of black versus white competence and reliability is a sensitive one. Suffice to say that perceptions do exist that Whites can be trusted to produce quality and to perform efficiently where Blacks cannot (ABC, 1989e). The existence of an acute backlog of housing in South Africa thus tends to favour the large (white) home building contractors when criteria such as speed of production and cheapness of the project are important (ABC, 1989a).

The attitude of black contractors towards white township developers is bitter. For example, Linda Nyembe, chairman of the then Professional Builders Federation (PBF) commented that "...the white builder is perceived to be taking the bread out of the black builder's mouth who is disadvantaged in a lack of adequate funds to develop and service the infrastructure of a new township" (ABC, 1989c). This bitterness is further reflected in allegations of corruption regarding the supply of housing contracts and land to white contractors. Many black contractors believe that white developers secure their work through corrupt relationships with officials (ABC, 1989c). On the acquisition of land, Creighton (1992), observed that :

> "... most of the land that was initially available for Black housing was within the boundaries of the existing Black townships and fell within the control of the newly established local Black authorities. Apart from the much criticised corruption suspected to be rife in the land allocation procedures which prevailed, few of the emerging Black building contractors were in a financial position to meet the terms and conditions of the allocation of stands."

Incompetent and unscrupulous contractors are difficult to identify prior to engaging them, particularly when they operate at the fringe of the industry. Consequently, there will always be those who stand to benefit from development assistance programmes when in fact they do not deserve to do so. This factor introduces the need for screening, monitoring and evaluation in order to "discriminate between small firms who genuinely need and respond to assistance and those who should not be helped" (ILO, 1987:35).

(b) Black expectations of assistance: The proliferation of organisations and programmes concerned with promoting the economic empowerment of black South Africans has had an effect on the expectations of Blacks. Several such programmes exist in the construction industry, which are generally aimed at uplifting black SSCEs, or improving their management capabilities. Merrifield (1992:65) observed that many black SSCEs who have participated in these programs have "developed expectations of receiving assistance regardless of their own contributions or capacities...".

Thus, the nature of this problem is that some black contractors have been lulled into complacency because of the degree to which black SSCEs, as a group, have been singled out for assistance or support. As a result, they neglect to develop their own coping mechanisms. The quality and availability of skilled and unskilled labour in the construction industry must also be seen in the context of raised Black expectations.

Sam Mabe, writing in the Sowetan newspaper, provided an important insight in this regard :

> "We also need to change our attitudes towards certain careers. A lot of vocational training centres in this country are half-empty because people look down on vocational training. We would rather be unemployed take training where we would have to use our hands and get dirty" (Mabe, 1989).

(c) Attitudes of Whites towards Black advancement: Commenting on a recent survey published by the South African Institute of Race Relations, Professor E. Webster noted that there were 63 Whites to eve one Black in managerial, executive and administrative positions in South Africa. The chairman of the National Manpower Commission, Dr. H. Reynders attributed this state of affairs to the attitude of Whi and concluded that they were either unwilling, or unable, to implement Black advancement programme (HSA, 1990a). This attitude is underpinned and perpetuated by the frequent poor performance of black businesses. In the literature, black leaders and businessmen themselves warn of the detrimental effect o attitudes towards Blacks in general of "sloppy work, bad planning and broken promises" (ABC, 1989e) addition, others have observed that black SSCEs "mismanage their businesses and misuse good opportunities" or that a builder "goes for luxurious things, spending the little profit he has made on fan cars..." (Padi, 1988c).

Considerable evidence surfaced in the literature that black contractors preferred building for white clien (ABC, 1988c; ABC, 1989a; ABC, 1990a; ABC, 1990d), because this enhanced their reputations. Refer to his rise in popularity, one builder commented: "Very amazing - people consider you capable only if can see you doing something for whites" (ABC, 1989b). There is, thus, an implicit perception that 'wh buildings' require higher levels of quality than township houses for Blacks. Involvement in 'white' wor appears to present a means whereby the black SSCE can improve its image.

2.5.4   A Case Study of a Small-Scale Construction Enterprise's Problems

The outline of the evolution of a particular South African black SSCE given in Box 2 is probably representative of many others. This summary traces the progression of A to Z Homebuilders - from su to failure and once again, to success.

2.6   INTERVENTION POLICY CONSIDERATIONS

The constraints described in the previous section represent a formidable challenge to those whose purpo is to alleviate them. They are so diverse that it is not immediately clear how, or indeed whether or not,

The firm was founded by two partners in February 1986. One partner had worked as a plastering sub-contractor for 10 years and then began general alterations and additions contracting, completing six jobs before joining A to Z. The other partner had been a bricklaying sub-contractor for a housing company prior to joining A to Z. The informal partnership commenced business on R2 000 of personal savings, operating from one partner's home. By April 1986 the firm had one project under construction and three pending. All were uncomplicated jobs. Just prior to completion of the first project the initial R2 000 had been spent and the firm had no means of financing its other commitments. The firm's application for an overdraft from a formal bank was unsuccessful.

The firm resorted to tendering on additional projects in order to raise working capital from the deposits. Due to their low overheads, they quoted "less per square metre than the average building contactor". They successfully secured one contract and used the deposit to finance and successfully complete their previous commitments. The firm then learnt that building material suppliers would extend credit if a cession of the client's mortgage bond could be arranged. With the cooperation of building societies and clients it obtained credit facilities with four suppliers.

After one year the firm appeared to be highly successful - six projects had been successfully completed, two were in the finishing stages and twelve were in progress. The combined value of this work was R330 000. In April 1987 a summons for R32 000 arrived from one of the suppliers - the partners had thought that they owed R5 000. In the ensuing negotiations, the supplier prepared an assessment of their overall position and found that: the firm had quoted below cost on sixteen jobs - a potential loss of R75 000; the firm stood to make a total R16 000 profit from four of its projects; R52 000 was required to complete work in progress, but only R35 000 was available for this from bonds; the firm owed material suppliers and sub-contractors a total of R58 000; and the partners combined assets were worth R6 000.

Acting on the advice of the supplier, the firm approached its creditors, offering to pay its debts over 5 years. All contracts were cancelled, but the firm still owed R43 500. The supplier then helped the firm obtain bridging finance from a bank and assisted it in quoting on new work. By December 1988, the firm had paid off all its creditors and was showing a before-tax profit of R30 000.

Source: Alexander (1990)

Box 2   A to Z Homebuilders - A Case Study

should intervene. This is essentially a question of "establishing priorities, identifying target groups, choosing among a variety of options and assigning resources. Even more difficult at the outset is to predict the cost-effectiveness of each of the options that might be available" (ILO, 1987:53).

The purpose of this section is to provide a contextual framework for the formulation of appropriate intervention policy in South Africa. Various organisations and institutions are in a position to influence, or implement strategies for, SSCE development. Such organisations exist in South Africa (see Appendix F) and, although some detail may be provided in this section, their effectiveness is separately analysed in Chapter 4. This section highlights certain prominent issues regarding intervention in SSCE development, as

reported in the ILO guidelines on the subject. In addition, certain suggestions arising from the local
literature, as to what might be conducive to the development of South African black SSCEs, are included
Five approaches were identified - (i) access to work opportunities could be improved, (ii) the formal
business environment could be made more favourable for SSCEs, (iii) access to training could be facilitat
(iv) national development agencies could be established and (v) existing SSCE representatives could be
used, or new such bodies established.

2.6.1 Improving Access to Work Opportunities

Government is usually the largest single client in the construction industry. As already noted, fluctuations
the supply of work from government have an important influence in the economy. In South Africa SSCE
are most likely to be affected by fluctuations in the availability of house building contracts.

Public sector investment in buildings in South Africa has been steadily decreasing over the last decade.
Gross domestic fixed investment attributable to the public sector in (i) all buildings, declined from 45% i
1980 to 36% in 1987 and in (ii) residential buildings, declined from 34% in 1980 to 19% in 1987
(Krafchik, 1990:212). As noted in section 2.3.3, this trend is set to change as South Africa moves towar
democratic government. The planning of public sector demand could greatly improve the prospects of
SSCEs gaining access to work. The ILO advocates a "rolling programme", which it defines as follows:

> "[t]he first step in the programme might be to publish indicative lists of projects in each sector, including th
> likely value and location. In the following year the list could be updated according to (a) progress in design
> and project planning, (b) any changes in priority, and (c) anticipated resources. At the beginning of each
> financial year a final list of projects could be released showing which were definitely to proceed in the cour
> of the year. ... To be useful the programme lists would have to be made available to small contractors at ea
> stage in the planning cycle." (ILO, 1987:56)

Another means of ensuring greater access by SSCEs to public sector work is the tender bias. This could
take the form of preference being offered to community-based contractors on a regional basis.
Alternatively, minimum values of public sector work for allocation to SSCEs could be determined and
enforced (ILO, 1987:58).

Tender prequalifications could be enforced to improve SSCEs' access to work. Generally, the objective o
tendering prequalifications "is to protect the client's interests in ensuring basic competence among
contractors bidding for work" (ILO, 1987:60). This could, however, be adapted to protect the small work
market by declaring larger contractors ineligible for contracts below a defined value and thereby ensuring
work opportunities for SSCEs.

It was argued in section 2.4.2 that the encouragement of sub-contracting would facilitate access to work
opportunities for a greater number of SSCEs than would otherwise be the case. Indeed, as has been show

South African SSCEs are largely ill-equipped to handle the demands of the general contracting environment. The conventional tendering process and the complexity of contracts pose serious problems for them. The simplification of documentation and the relaxation of pedantic controls could therefore considerably enhance the likelihood of SSCEs bidding for work (ILO, 1987:69).

The intervention of government in the private sector market can have an adverse effect on competition. Whether or not to intervene must be contemplated in the light of public accountability (ILO, 1987:69). However, government could play a relatively neutral role in helping SSCEs gain access to work by means of financial inducements to private sector clients. In South Africa, the government could do much to create a market for small developers and contractors by ensuring that they have access to development land (Padi, 1988c; HSA, 1991c).

2.6.2   Creating a More Favourable Business Environment

The creation of a more favourable business environment for SSCEs is something that both the public and the private sectors could be involved in. It is common for clients to require performance guarantees from contractors. The guarantee is usually for 10% of the contract amount and serves to protect the client in the event of the contractor failing to complete the contract. SSCEs often experience difficulty in obtaining these guarantees and consequently are effectively denied access to certain work. However, the ILO argues that "it is rare for performance bonds to be called in", despite the fact that SSCEs often fail to complete contracts (ILO, 1987:71). This suggests that their usefulness is questionable. Their main effective use might thus be to filter out financially dubious SSCEs. However, given that tendering prequalifications can effectively be used to assess whether or not SSCEs are risky, the abandonment or relaxation of performance guarantees would contribute to improving the business environment of SSCEs (ILO, 1987:72).

We have seen that the lack of access to loans is a significant obstacle for SSCEs. This problem can be overcome through the establishment of loan guarantee funds which ensure that SSCEs have access to loans. The ILO's experience shows that either partial or full guarantees can be offered with commercial banks and/or parastatal development corporations administering the scheme. The Kenyan experience, where credit guarantees were extended to African SSCEs through a development corporation was discouraging. Repayment records were dismal, but this appeared to be largely the result of poor initial screening. Based on the Kenyan problems, the ILO has suggested the following guidelines:

> "(i) *not all small contractors should be eligible* : there must be some attempt to identify a target group of contractors who meet eligibility criteria in terms of location; size of business; proven commitment to construction; and minimum basic competence. ... Financial irresponsibility should be grounds for their deletion from the target group;
>
> (ii) *loan application should be screened and appraised* : some agency should be made responsible for assessing whether a loan will be properly used, that it is of an appropriate amount, and that the contract on which it will

be used is itself likely to be financially viable;

In South Africa the Urban Foundation (UF) has established a loan guarantee fund, but this is intended to provide access to mortgage bonds for home buyers, mainly in the under R35 000 market (HSA, 1989d). The issue of access to finance is one which involves both the builder and the potential home owner. Without mortgage bonds, individuals cannot commission the building of houses. Thus, if their potential clients do not have access to mortgage finance, the SSCEs' supply of work diminishes, as does their access to the bridging finance arrangements which are often secured against these bonds. Black home owners have earned a poor reputation with the formal lending institutions through the practice of boycotting bond repayments. The result of this has been a reluctance on the part of these institutions to lend to black home buyers. And this has been detrimental to the contractors who operate in those markets.

The introduction of banking codes of practice should be considered in the interests of improving communication between bankers and unsophisticated SSCEs. It is often the case that SSCEs initially approach, and later use, the banking system in ways that do little to inspire the banks' confidence in them. The ILO suggests that bankers could do much to improve this situation by specifically making SSCEs aware of what would be considered acceptable or favourable. This could be done by publishing information brochures explaining: interest rate structures, collateral requirements and reasons why and how these apply in various circumstances; how they expect business and personal banking accounts to be used; and how they expect loan applications to be compiled and presented. Further, loan officers are often unfamiliar with the idiosyncrasies of the construction environment and banks could ensure that suitably trained personnel were available to process contractors' loans (ILO, 1987:77).

We have seen from Merrifield's (1992) research that SSCEs are severely disadvantaged by their lack of access to economies of scale. It is essentially only the large developers who manage to benefit from bulk buying. Further, the tendency is for these developers to issue sub-contract work to smaller firms on a labour-only basis. In this way the developer still benefits from material discounts which the sub-contractor would not have received.

Considering the rapid and regular increases in building material prices, access to discounts is becoming increasingly more important for the success of contracting firms in South Africa. Krafchik (1990:180) has called for further research into the question of why building material prices continue to rise steeply -

Certain programmes, for example, SAHT buy materials in bulk on behalf of SSCEs, but these materials are not available to SSCEs in general (ABC, 1989a).

The process of white and black contractors entering into joint ventures has been argued by some to be the only way in which black SSCEs can rapidly rise to the level of sophistication required in the contracting market (ABC, 1989d; ABC, 1990d; HSA, 1989a). Policies or mechanisms which encouraged white firms to enter into such ventures could significantly assist the development of black SSCEs. This idea can, indeed, be expanded to include established and successful black firms entering into joint ventures with less experienced and capable firms. There is some evidence that this occurs in the black contracting community (ABC, 1988d), but the practice is unlikely to be widespread.

2.6.3 Providing Training and Technical Advice

Many of the problems experienced by black SSCEs stem from a lack of: schooling; skills training; or management training. It is then clear that, the broader the base of the training infrastructure, the more beneficial it will be for SSCEs. It has already been established that the training available in the building industry is largely inaccessible to most black SSCEs. Clearly, this situation will have to be remedied. This section thus summarises the factors that the ILO (1987) has found to be important in initiatives aimed at instituting training for SSCEs. These factors can be presented in the form of the following questions:

(a) Where should training take place? : Consideration must be given to the geographical distribution of contracting firms as this will help decide whether it would be more appropriate to expect contractors to travel to training centres or to devise means of delivering on-the-job-training to the firms (ILO, 1987:84).

(b) When should training occur? : Active contractors are only likely to be able to attend training sessions over weekends and in the evenings. Training courses need to be structured so as to make this possible (ILO, 1987:84).

(c) What should contractors be trained in? : It is important for trainees to acquire at least some new skills early on in the programme. Further, these should ideally be able to be put to immediate use. The focus should therefore be on improving the major problem areas that contractors experience such as: preparing loan applications; computing working capital requirements; and compiling tenders (ILO, 1987:85)

(d) Who should train? : It is fundamental to the success of the training programme that the right trainers are used. Trainers must have been taught how to train effectively and, above all, must have a very clear idea of what it is like to be a small contractor (ILO, 1987:85).

Training publications can play a vital role in disseminating sound skills and management practices. SSCEs

tend to learn more effectively when instructional material is interspersed with case studies and examples
based on common local procedures and problems (ILO, 1987:90). The journal, African Building Contract
has frequently published such instructional articles, but it would need to reach a significantly larger
subscription market before it could be considered an effective training instrument.

2.6.4   Contractor Development Agencies

The ILO contends that the most pragmatic method of providing assistance to SSCEs is through the
establishment of a contractor development agency (CDA) under government auspices (ILO, 1987:95). Th
section does not deal with how best to establish CDAs, but generically summarises the types of assistance
which the ILO believes should be administered through them.

CDAs can provide five main types of support: (i) flow of work; (ii) training and advisory services; (iii)
finance; (iv) corporate approach; and (v) physical support" (Edmonds and Miles, 1984:117;
ILO, 1987:100).

One manner of ensuring a flow of work for SSCEs is for the CDA itself to act as an entrepreneur,
tendering on large contracts and then distributing the work among its registered members. The ILO advise
caution in adopting this approach and reports that existing CDAs in Africa do not generally take on the ro
of contractors (ILO, 1987:100).

The dissemination of training and advice to its members is the most important role the CDA is likely to
play. ILO research shows that the effectiveness of training courses varies widely, but that the best results
are achieved when training is tailored to the particular needs of groups of contractors. Further, the conten
of training courses is much more likely to be attractive to potential trainees when they themselves have be
consulted on what would be appropriate course content (ILO, 1987:101). Advisory services often need to
capable of being mobile. When advice concerning site operations is required by contractors it is often
essential that the advice can be given on site, rather than in the form of literature, *etc.* (ILO, 1987:104).
is essential that legal assistance and advice be made available through CDAs. This would need to cover
specific contractual issues as well as general aspects such as insolvency, *etc.* (ILO, 1987:105).

SSCEs generally regard the financial services offered by CDAs as the most beneficial. Types of financial
services could include: short term loans (working capital); sureties in respect of performance guarantees,
etc.; insurance sureties; and medium-term investment loans (ILO, 1987:106).

A national CDA would be in a good position to act as a channel of communication between government a
SSCEs. The CDA could play the role of monitoring changes in policy, standards and prescribed practices
and keep SSCEs informed of these. Further, the CDA could act as a useful channel to communicate the

problems and needs of SSCEs to the various legislative bodies (ILO, 1987:111).

The provision of plant and equipment is a further important function which the CDA could perform. SSCEs would clearly benefit enormously by having reasonably easy access to hired equipment, vehicles, *etc.*, and instruction on how and when to use various pieces of equipment (ILO, 1987:112).

Although no national government sponsored CDA exists in South Africa, contractor development agencies and programmes are currently active. These organisations, their activities and their effectiveness are discussed in Chapter 4 and Appendix F.

While it would be interesting to present a discussion on how the attempts to set up CDAs elsewhere in Africa turned out, it would exceed the scope of this dissertation. What must, however, be questioned briefly, is whether or not the Kenyan National Construction Corporation (NCC) (probably the best known of the CDAs) was a success. There are mixed opinions about this. Wells (1986:122) voices the concern that after 15 years:

> "the most tangible achievement of the NCC is that, through the provision of loans and its intervention as an intermediary in the tender market, it helped a small number of African contracting firms to capture a share of public-sector building work and a less significant share of civil-engineering work....African firms are still undertaking only the very smallest and least sophisticated of contracts. And their share of the total construction market is still miniscule *(sic)*"

Edmonds and Miles (1984:130), on the other hand take a much more positive view of matters, believing that,

> "despite its problems, the NCC remains the most comprehensive and sustained example of a contractor support agency in Africa (and probably elsewhere)...The NCC, as a pioneer, has taught useful lessons on both the possibilities and the pitfalls. It has bravely attempted to grasp the former and sidestep the latter."

Wells (1986:111) suggests that part of the problem with the NCC might have been that SSCEs regarded it more as a permanent source of work and finance, than as an interim phase on the road to self-reliance. Why would they want to become independent of the NCC under those conditions? There are obvious reasons why a contractor would want to stay dependent on the NCC. Merrifield (1992:64) makes the same observation about South African contractor development programmes, noting that the establishment of these programmes does not necessarily ensure that they will achieve their objectives. In this regard, he comments:

> "A number of black builders I spoke to however expressed dissatisfaction with those programs. In particular, they felt that the programs were too tightly controlled and that they had little scope for exercising their own initiative. ... They implied that there was little difference working for these programmes and for working for any other private sector developer.
>
> I would like to express the concern, on the basis of my limited exposure to these programmes, that

programmes designed to uplift the small black builder run the risk of creating a culture of dependency which
could ultimately hold back the development of black contractors."

It is important to consider that, without some form of assistance SSCEs in general, and black SSCEs in
particular, are unlikely to develop into MSEs and LSEs (Van Staden, 1993; Watermeyer, 1993). Somethi
must clearly be done, but what? This section has noted that CDAs could provide important parts of that
assistance, but also that CDAs will not necessarily succeed at what they set out to accomplish. It is
possible, though, to amend the theory of what should work in light of what did not work. Since future
government intervention in SSCE development can be expected, the concept of the CDA is important. Th
recommendations of this dissertation are thus likely to be influenced accordingly.

2.6.5   Contractors' Associations

The role of contractors' associations (CAs) is a potentially important one (ILO, 1987). Not only can
governments and the private sector be of assistance to developing SSCEs, but so can the SSCEs themselve
However, a precondition for this is that they are unified as a federation or association (ILO, 1987:121).
Herein lies a problem. SSCEs do not appear to see the benefit in this and consider it illogical to join force
with their competitors (ILO, 1987:121). This section summarises the ILO view on how CAs can help
SSCEs.

Firstly, CAs can form a link whereby SSCEs can present their problems to government, that is, external
representation. But, for this to be effective government's response would have to be organised. This coul
be achieved through the establishment of "specifically appointed and authoritative representatives of
government" (ILO, 1987:122). In this forum, CAs could discuss policy issues such as "materials prices a
supplies, cost escalation, tendering procedures, interim payments, the provision of training facilities for
skilled entrants to the industry..." (ILO, 1987:122). The representation of SSCE interests can, of course,
extended beyond government. CAs could interface with "training institutions, materials suppliers, the
insurance industry, professional consultants..., customers of the industry, commercial bankers [and]... the
industry's work force" (ILO, 1987:122). Indeed, in South Africa, financial institutions are reportedly mor
willing to consider loan applications from firms belonging to CAs (Mngomezulu, 1988). In addition, it w
also evident that CAs had been in a position to acquire development land and distribute it among members
(Padi, 1989a).

Secondly, CAs can provide internal services such as "information dissemination, advice, training and
administration". Such services have proved to be valuable contributions by CAs to the support of SSCEs
(ILO, 1987:124). Problems arise when the CAs get too large, or represent too diverse a range of
contractors. Sometimes, the smallest contractors are simply refused membership, while in other instances,
they are neglected while the larger firms dominate the internal services (ILO, 1987:124). In the South
African literature there was evidence that, to some extent, SSCEs have indeed obtained training, assistanc

and advice from their CAs (Padi, 1989b; Padi, 1990b; Padi, 1990c), but this does not appear to occur on a large scale.

Finally, CAs could play a major role in setting standards (ILO, 1987:124). The ILO suggests that SSCEs could benefit by holding certificates of competency issued by their CAs. This would help in procuring work (ILO, 1987:124). In essence, CAs could compile graded prequalification lists. For such a system to be effective, competency and compliance with standards would have to be policed by the CAs.

Over and above these three possible functions of CAs, the ILO (1987:125) suggests that they could play a more radical role in developing countries where no formal or widespread SSCE support action has yet been taken. They could perform some of the functions and services of a CDA and indeed as will be noted in Appendix G, certain local CAs recognise this.

2.7   SUMMARY

This chapter has reviewed the literature on SSCEs and issues related to their development. Development strategies, whether implemented by the public or private sector will always be influenced by the overarching socio-economic policies of the state. For this reason sections 2.1 and 2.2 highlighted the issues regarding likely shifts in current socio-economic policies and the impact of these on the construction sector of the economy. Given the tendency for small firms to undertake small works such as houses, section 2.3 was included in an attempt to broaden the common myopic perception that 'housing' is 'houses'. Indeed, it is expected that this section consolidated the concept that 'housing' is a process. It therefore includes those who build houses, and state housing policies therefore affect small contractors.

With the preceding sections having identified a role for SSCEs in the delivery of housing, sections 2.4 and 2.5 brought into focus the issues of what SSCEs are and what factors are known to hamper their growth and development. Section 2.6 took this further by reviewing possible intervention strategies for the development of SSCEs. Apart from the studies of Krafchik (1990) and Merrifield (1992) which were limited, respectively, in scope and depth, the literature concerned SSCEs in foreign countries. As will be noted in Chapters 5 and 7, much of the detail in the international literature is applicable to South African SSCEs. However, it is obvious that intervention policies for South Africa would be ill-conceived if they were insensitive to the particular attributes and problems of South African SSCEs. This issue is addressed by the depth of the questionnaire.

CHAPTER 3

AN OVERVIEW OF THE SOUTH AFRICAN CONSTRUCTION SECTOR

3.1 INTRODUCTION

This chapter provides an overview of the construction sector as reflected in the official South African economic statistics. Since the commencement of intervention in SSCE development is likely to precede a major economic restructuring in South Africa, this chapter has been included to facilitate an understanding of the current structure of the construction sector of the economy. Further, since it is expected that the majority of black SSCEs would operate in the homebuilding sub-sector, emphasis is placed on the building industry and the activities of Blacks in the construction sector.

It must be noted, however, that the official statistics exclude government work. The data also exclude the value of work, wages and salaries and employment attributable to informal contractors. However, in the case of informal sub-contractors, value of work is included in the value of the principal contractor's work but wages and salaries and employment are excluded (Krafchik, 1990:71). For these reasons, the size of construction sector is likely to be considerably understated in the descriptive profile presented in this chapter.

3.2 DEFINITION OF THE SECTOR AND SUB-SECTORS

3.2.1 Elements of the Construction Sector

In the most recent Census of Construction (1985), the construction sector is sub-divided into the following elements ('a' to 'i' below) :

THE CONSTRUCTION SECTOR

	(a)	Home builders
	(b)	Other building construction by general contractors
	(c)	Painting and decorating
BUILDING INDUSTRY	(d)	Plumbing
	(e)	Electrical Contracting
	(f)	Shopfitting
	(g)	Other special contracting, not elsewhere classified
	(h)	Civil engineering
CIVIL ENGINEERING INDUSTRY	(i)	Construction, not elsewhere classified

For the purposes of this study items (a) to (g) are collectively defined as the 'building industry' while items (h) and (i) are defined as the 'civil engineering industry'. Together they make up the 'construction industry'.

3.2.2 Definition of the Building Industry

The definition of the building industry is contentious. A broad definition is provided by the World Bank (1984:29) - "Building construction usually includes housing, factory shells, office buildings, schools, hospitals, barracks, and farm buildings." However, in this study, the following definition, provided by the Industrial Council Agreement for the Cape Peninsula, is adopted :

> "'Building Industry' means ... the Industry in which the employer and the employee are associated for the purposes of erecting, completing, renovating, repairing, maintaining or altering buildings and structures *and/or making articles for use in the erection, completion or alteration* of buildings and structures, whether the work is performed, the material prepared or the necessary articles are made on the sites of the buildings or structures or elsewhere, and shall include all work executed or carried out by persons therein who are engaged in the following activities or sub-divisions thereof, including excavations and the preparation of sites for buildings as well as the demolition of buildings, unless it can be shown by the employer concerned that such demolition was not carried out for the purpose of preparing the sites for building operations : Bricklaying...; french polishing...; joinery...; leadlight-making...; masonry...; metal work...; painting...; plastering...; plumbing...; shop, office and bank fitting...; steel reinforcing...; steel construction...; woodworking..." (Republic of South Africa, 1987c).

This definition lacks clarity regarding the inclusion of manufacturers in the building industry. Indeed, the phrase "*and/or making articles for use in the erection,..*" suggests that a variety of manufacturers and retailers could conceivably be part of the building industry. Turning to the BIFSA definition of the building industry - which is almost identical to the above definition - it is clear that there are circumstances under which manufacturers and retailers should not be included. This definition clarifies the issue by the inclusion of the following qualification : "...provided, however, that any person who manufactures such articles *for sale purely* shall not to that extent and for that reason alone be regarded as being in the building industry..." (BIFSA Official Handbook).

3.2.3 Definition of the Civil Engineering Industry

The broad definition given in a World Bank (1984:29) study states that - "Civil engineering construction includes highways, water supply, power generation and irrigation structures, airports, railways, ports, and the like." Central Statistical Services, Pretoria, provides the following definition of the civil engineering industry in its specification for categories (h) and (i) (see section 3.1). This latter definition is adopted for the purposes of this study :

> "*Contractors engaged in* constructing, altering, and repairing roads and streets and bridges; viaducts, culverts, sewers, and water, gas and electricity mains; railway road-beds, subways, harbours and water-ways; piers, airports and parking-areas; dams' drainage, irrigation, flood-control and waterpower projects and hydro-

electric plants; pipe lines; water wells; athletic fields, golf courses, swimming pools and tennis courts;
communication systems such as telephone and telegraph lines; marine construction, such as dredging and
under-water rock removal, pile-driving, land draining and reclamation; and other types of heavy construction
Businesses primarily engaged in performing mining services such as preparing and constructing mining sites
shaftsinking, underground cementation and drilling crude oil and natural gas wells, on a contract of free basi
are classified in this group."

3.3 THE CONSTRUCTION SECTOR IN THE SOUTH AFRICAN ECONOMY

3.3.1 Contribution to Gross Domestic Product

Research in developing countries has shown that the construction sector contributes between 3 and 8 per
cent to GDP. The size of construction's contribution tends to increase as a country's resource base grows
levelling out once the economy is highly developed (World Bank, 1984:11). This tendency has, indeed,
been evident in South Africa where the contribution to GDP of the construction sector rose from around

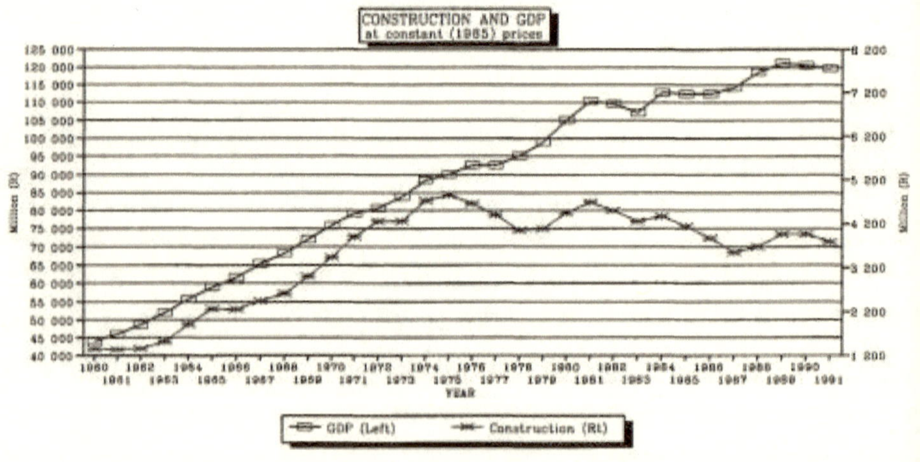

Source : SA Statistics (1992, 21.6)
Figure 3.1 - Real Value of GDP and Contribution of Construction Sector

in 1960 to 5,2% in the mid seventies. This can be seen in Figure 3.1 where it is also clear that after
decreasing slightly from the 1974-75 peak, construction has levelled out at about 3,3% since 1986 (also se
Table C1 - Appendix C).

Examining growth of both construction and total GDP, it can be seen that construction generally mirrors
total GDP, but experiences greater volatility (see Figure 3.2). Further, construction has tended to
experience negative growth more frequently than GDP. This tends to support frequently cited findings (fo

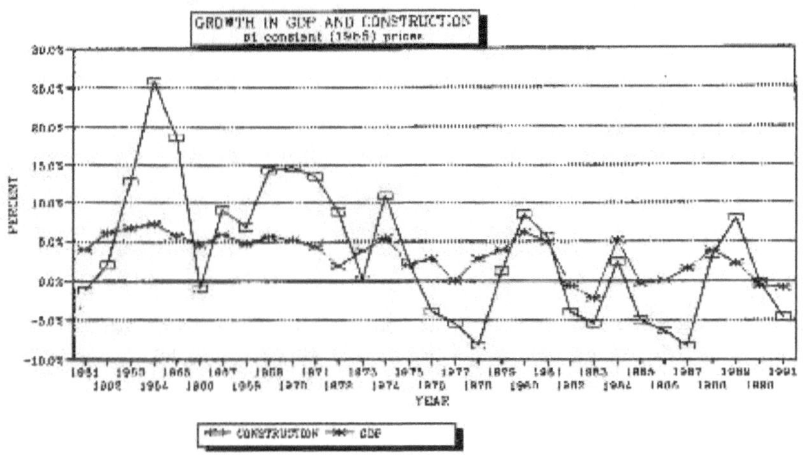

Source : SA Statistics (1992:21.6)
Figure 3.2 - Historical Growth of Real GDP and Contribution of Construction Sector

example, World Bank (1984:39)) that the construction industry fluctuates more widely than other sectors.

3.3.2   Contribution to Non-Agricultural Employment

The contribution of the construction sector to total non-agricultural employment (excluding working proprietors) is modest. Table C2 (see Appendix C) tabulates data on employment and growth in employment for all non-agricultural sectors and for construction. Based on this table, the share of the construction sector in total employment is reflected in Figure 3.3.

The relatively greater volatility of construction industry output, referred to above, is mirrored when growth in total employment and construction employment prior to 1983 are compared (see Figure 3.4). It is interesting to note the convergence of growth rates after 1983. This may reflect a tendency towards: smaller casual and more stable permanent work forces in formal firms; and sub-contracting to informal firms where hiring and firing is not recorded in official statistics.

The distribution of employees across race groups is shown in Figure 3.5. An analysis of the growth rate of the various race groups reveals that employment levels tend to fluctuate very similarly for all race groups. It is noteworthy that numbers of black employees have tended to increase and decrease more steeply than employee numbers in other race groups (see Table C3 - Appendix C). This is related to the greater use of black workers and unskilled casual workers, with less job security. Retrenchment has traditionally been widespread in the construction industry - in 1990 the level of retrenchment in the construction sector was six times that of retail and three times that of manufacturing (BIFSA 1990:48). Further, retrenchment is

58

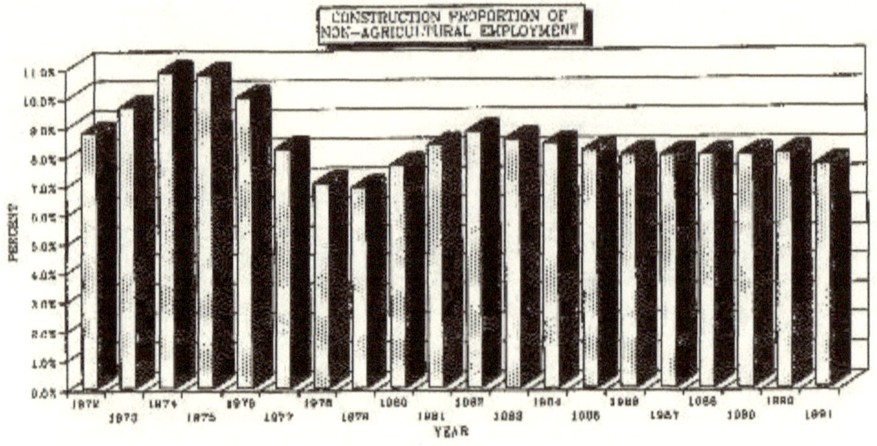

Source : SA Statistics (1992: 7.8-7.13)

Figure 3.3 - Contribution of Construction Sector to Total Non-Agricultural Employment

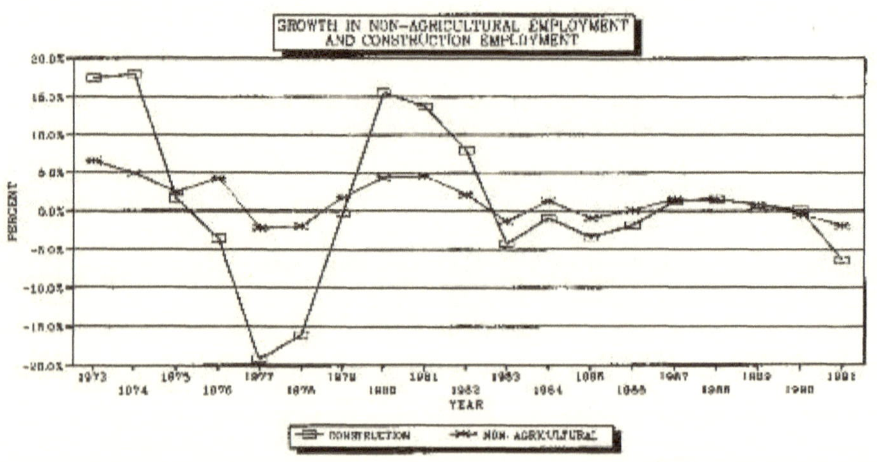

Source : SA Statistics (1992: 7.8-7.13)

Figure 3.4 - Annual Growth in Non-Agricultural Employment and Construction Sector Employment

increasingly being targeted by trade unions (BIFSA, 1990:47). This is likely to hasten the growth of labour only sub-contracting and should reflect as a greater conformity in the growth of construction employment and total non-agricultural employment.

Figure 3.6 indicates that the number of working proprietors in the construction industry increased between 1972 and 1985. Given that the real value of construction work has remained fairly constant since 1972 (see Figure 3.1), this seems to indicate a tendency towards smaller firms. Indeed, Krafchik's (1990:71) findings

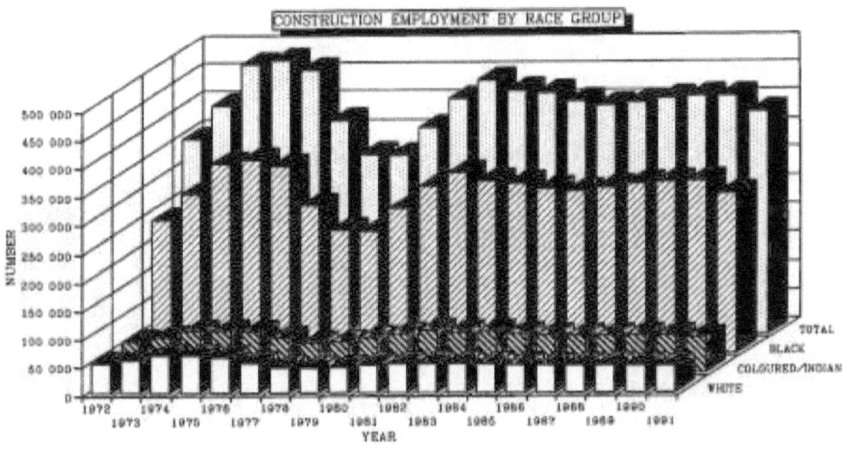

Source : SA Statistics (1992: 7.13)

Figure 3.5 - Total Construction Sector Employment and Distribution by Race Group

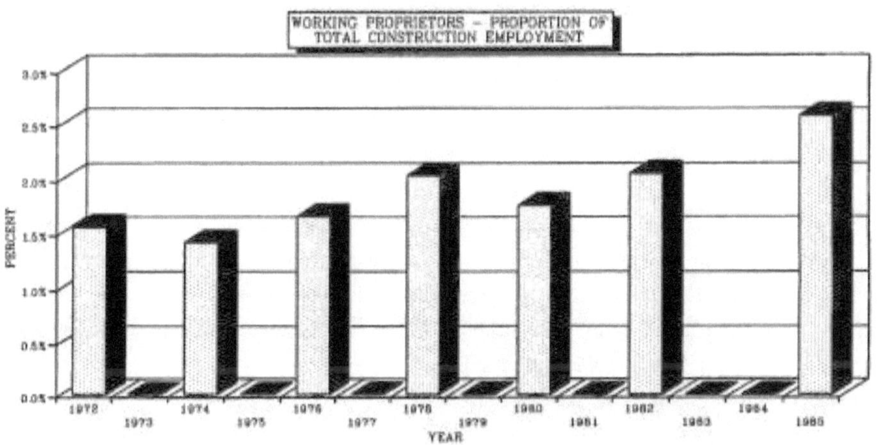

Sources: Census of Construction (1985:1)

Figure 3.6 - Working Proprietors - % of All Construction Sector Employment

confirm this - construction firms "with fewer than 20 employees, as a proportion of all establishments, increased from 52.6% in 1966 to 75.3% in 1985".

3.3.3   Contribution to Total Wages and Salaries

Wages and salaries paid to construction employees (excluding working proprietors) amounted to R4.7

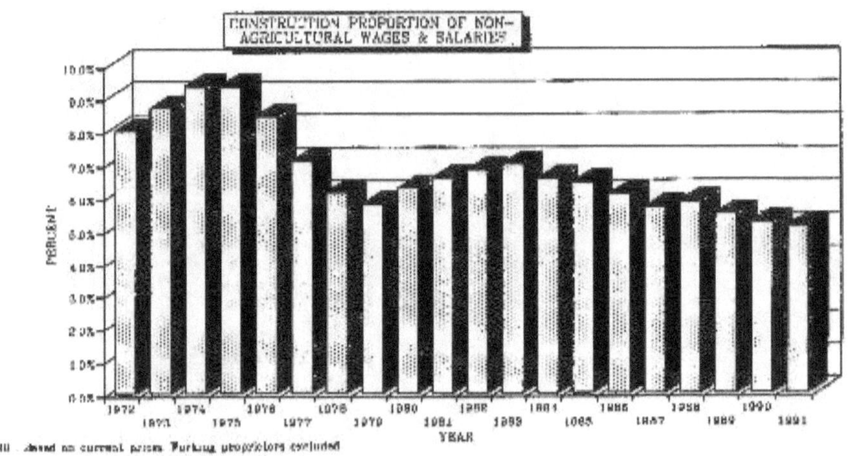

Figure 3.7 - Construction Proportion of Total Non-Agricultural Wages & Salaries

billion in 1989 (see Table C4 - Appendix C). After 1974, construction wages and salaries declined stead
as a proportion of total non-agricultural wages and salaries (see Figure 3.7). Indeed, it is expected that t
would mirror the trends in construction value and employment (observed in Figures 1 and 3). Similarly,
annual growth in construction wages and salaries (see Figure 3.8) mirrors growth in construction value
Figure 3.2), tending to be more volatile than growth in total wages and salaries.

An examination of trends in the value and growth of construction wages and salaries by race group reve
that prior to 1979, the total paid to whites generally exceeded the total paid to Blacks (see Table C5 -
Appendix C). Since 1979, however, black wages and salaries have exceeded that of whites and growth h
generally been faster (see Figure 3.9). Notably, 1979 was the first year when Blacks were allowed to fo
and join trade unions. These data indicate that Blacks in the construction sector have been increasing in
importance. This is related to the spread of unionisation.

Figures C1 to C4 (Appendix C) compare average monthly salaries for construction, in total and by race
group, with the corresponding data for total non-agricultural sectors. From Figure C1 it is evident that
average monthly wages and salaries in the construction sector have lagged behind the average for total n
agricultural employment - as at 1989 construction had not yet penetrated the R1 000 per month level.

Similarly, an analysis by race group reveals that average wages and salaries paid to black construction
sector employees have generally been declining in relation to the average for all non-agricultural Blacks.
can be clearly observed that, up until 1977, construction Blacks earned more on average than all non-
agricultural Blacks (see Figure C2), but since 1978 have remained below the average. In contrast, it can
seen from Figures C3 and C4 that the white and coloured/Indian race groups have consistently earned ab

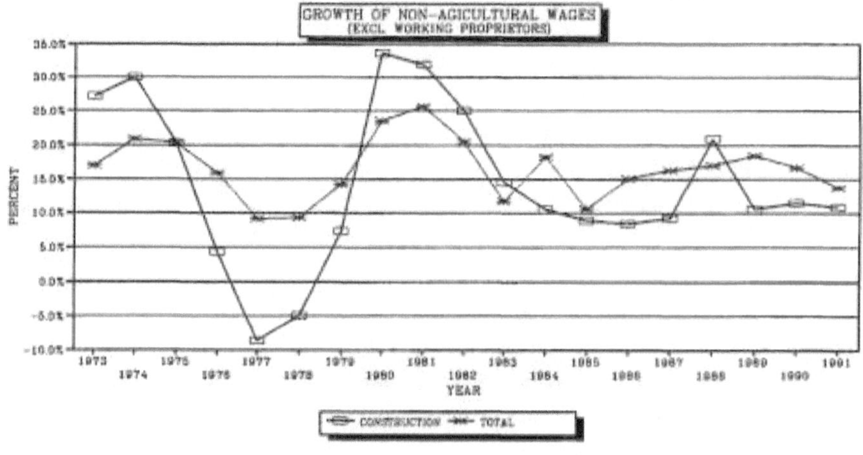

Source : SA Statistics (1992: 7.8-7.13)

Figure 3.8 - Growth in Wages & Salaries - Construction and Total Non-Agricultural

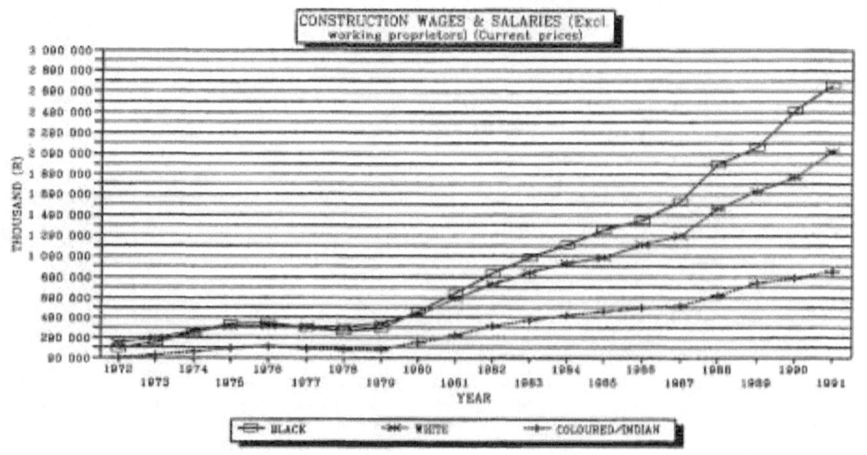

Source : SA Statistics (1992: 7.8-7.13)

Figure 3.9 - Distribution of Construction Wages and Salaries by Race Group

the averages of their respective race groups (see also Tables C6 and C7 - Appendix C).

The decline in black wages and salaries must be considered against the background of the ongoing shortage of skilled labour in the construction sector (Krafchik, 1990:145) and the oversupply of unskilled labour. Whites, Coloureds and Indians have a monopoly over construction industry skills and due to the general shortage of skills, this tends to strengthen their bargaining power relative to the Black race group. Further, the relative decline in average black wages and salaries appears to reflect a lesser degree of unionisation among construction Blacks relative to Whites, Coloureds and Indians.

Nevertheless, BIFSA (1990) predicted that black labour would become increasingly more politicised. Further, BIFSA (1990:42) reported that almost all employers of more than 200 individuals (in all sectors) had been organised by the unions. The focus is now on the smaller companies. In this regard, 40% of strikes in 1989 (in all sectors) involved firms with fewer than 200 workers (BIFSA, 1990:42) - two-thirds of them being over the wage issue (BIFSA, 1990:45).

3.3.4   Expenditure on Gross Domestic Product

The importance of the construction sector in the economy lies in its contribution to gross fixed investment. World Bank (1984:11) data indicate that the value of construction usually accounts for more than one-half of gross fixed investment in developing countries. Levels of fixed investment in the South African economy are similarly high.

Figure 3.10 shows total (public and private sector) investment in buildings and construction in the South African economy. In 1989 investment in buildings and construction represented 44% of total investment.

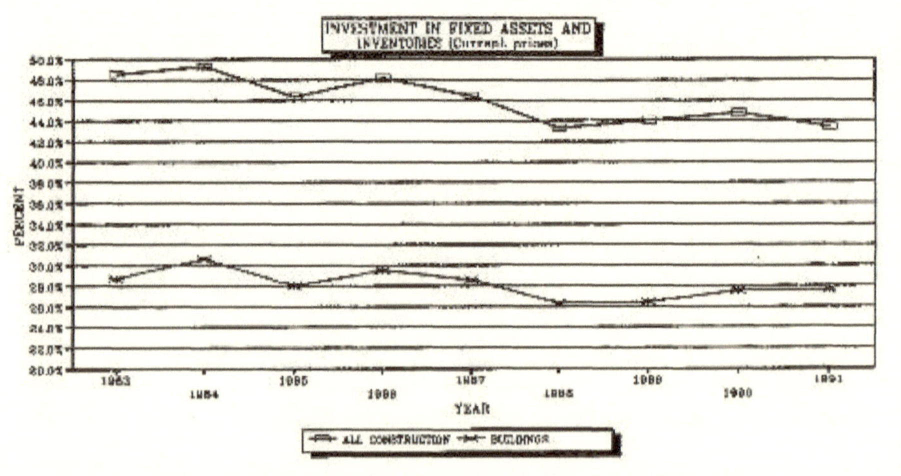

Source : BIFSA (1983 - 1987; 1988a; 1989 - 1992)
Figure 3.10 - Investment in Fixed Assets & Inventories - Construction and Buildings

Considered in isolation, investment in buildings formed 26% of total investment in 1989 (refer to section 3.4.3 for more detail). Notably these levels tended to reduce slightly between 1983 and 1989, dropping more sharply after 1986 (see also Table C8 - Appendix C). This might be related to declining public sector investment in new buildings (see details in section 3.4.3) and political uncertainty/instability.

3.4 THE BUILDING INDUSTRY IN THE SOUTH AFRICAN ECONOMY

This section considers the role of the building industry and its sub-divisions in the construction industry. For the purposes of this discussion, the building industry can be divided into the following divisions :

(i)   Home builders (ii)   General contractors   (iii)   Specialised contractors

3.4.1 Employment in the Building Industry

(a) Building industry share of total construction employment (inclusive of working proprietors) : As could be expected from the relative sizes of the building and civil engineering industries, the building industry employs the majority of construction industry employees. Table C9 (see Appendix C) clearly reflects this. Working proprietors in the construction industry are almost all - in excess of 90% since 1976 - employed in the building industry (see also Table C11 - Appendix C). This again reflects the small size of firms in the building industry as seen in section 3.2.2.

(b) Homebuilding division of the building industry - share of building industry employment (inclusive of working proprietors): Homebuilding has only been reflected as a separate entity in the South African statistics since 1982. Prior to that, homebuilding and general contracting were given together. Figure 3.11 reveals that the proportion of building industry employees representing the homebuilding division declined between 1982 and 1985 (see also Tables C10 and C13 - Appendix C). There was, however, a relative increase in the number of working proprietors in this division. This could be related to: the exodus of large firms from the homebuilding market and the general tendency towards smaller firms; and perhaps the greater use of informal sub-contractors. The tendency towards smaller firms is reported by Krafchik (1990:75) to be a feature of the homebuilding industry - 80% of homebuilding firms have fewer than 20 employees.

(c) Specialised contracting division of the building industry - share of building industry employment (inclusive of working proprietors) : The specialised contracting division of the building industry consists of those trades which are *usually* sub-contracted out by principal contractors. However, these specialist firms often operate in the capacity of principal contractors and trends in this division do therefore not necessarily reflect trends in sub-contracting.

Working proprietors and paid employees in this division increased their respective shares relative to building industry employment over the 1972-1985 period (see Tables C10 and C14 - Appendix C). In this regard, it is interesting to note Krafchik's report (1990:71) that "the extent of sub-contracting in the industry has...increased significantly...35.1% of gross output in the construction industry in 1972 [and] by 1985 this proportion had risen to 45%...". This finding suggests that the data in Figure 3.11 do, indeed, reflect an increase in sub-contracting.

(d) General contracting division of the building industry - share of building industry employment (inclusive of working proprietors): The general contracting division of the building industry contributes a very small number of working proprietors to total building industry numbers of working proprietors (see Tables C10 and C12 - Appendix C). This is related to the concentration of small firm working proprietors in the special contracting and homebuilding divisions. It is important to remember that these data probably understate numbers of working proprietors in general contracting due to the exclusion of informal general contractors from census data.

Figure 3.11 indicates that total working proprietors and paid employees in the division have increased in relation to building industry employment. This could be related to the exodus of working proprietors from the homebuilding division and their entrance into general contracting.

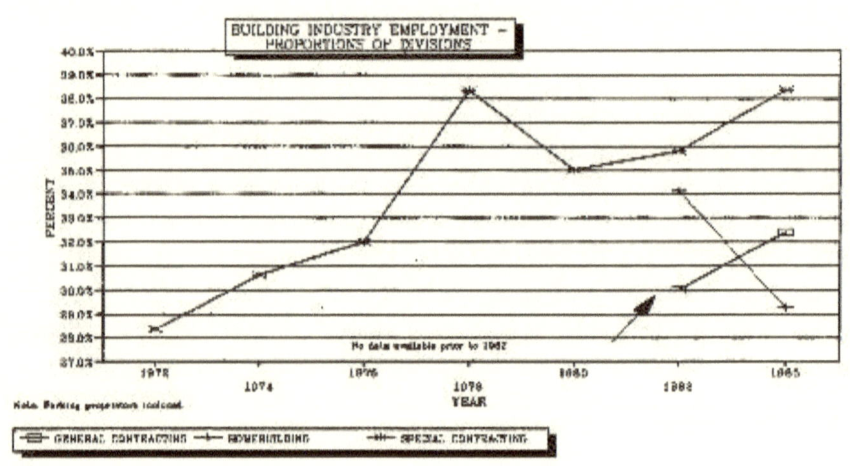

Source : *Census of Construction (1985: 1-2)*

Figure 3.11 - Building Industry Employment - Contribution of Divisions

3.4.2 Gross Output

(a) Total building industry share of gross construction output : In a World Bank (1984:30) study of nineteen countries, it was established that the contribution of the civil engineering and building industries gross construction output is generally 30% and 70% respectively. The report further states that in

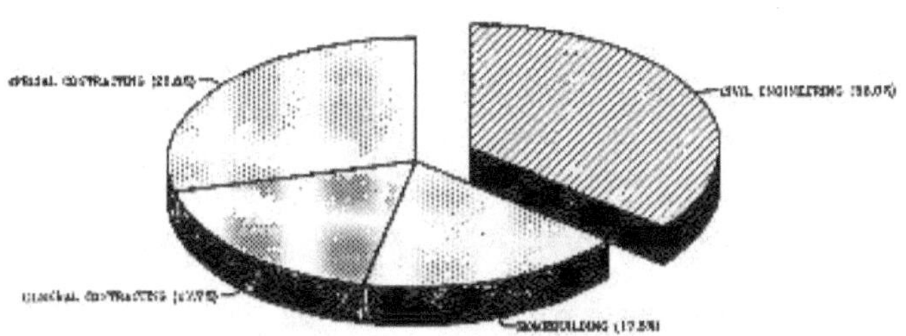

Source : Census of Construction (1985: 1-2)
Figure 3.12 - Construction Gross Output - Contribution of Civil and Building Industries

developing countries, where the informal sector plays a significant role in the supply of housing, the proportion of building is likely to be greater than official statistics show. South African data is largely consistent with this (see Tables C15 and C16 - Appendix C, and Figure 3.12). It must be noted that there has been a decline in the share of building industry contribution to construction gross output over the last two decades. This might be attributable to greater unmeasured informal contracting and sub-contracting in building relative to civil engineering.

(b) Homebuilding division of the building industry - share of gross building industry output :
Homebuilding is responsible for almost one-third of building industry gross output (see Tables C15 and C16 - Appendix C, and Figure 3.13). The decline in gross output between 1982 and 1985 mirrors the decline in employment in this division as seen in section 3.4.1 (b) above.

(c) Specialised contracting division of the building industry - share of gross building industry output : In line with the trend in employment in this division, the contribution of the specialised contracting division to building industry gross output increased between 1972 and 1985 (see Tables C15 and C16 - Appendix C, and Figure 3.13). Apart from any increase in sub-contracting, it must be noted that, in their capacities as principal contractors, painting and electrical contractors in this division could have undertaken increasing volumes of repainting, rewiring and other such work over this period.

(d) General contracting division of the building industry - share of gross building industry output : It can be seen from Figure 3.13 that the relative share of gross building industry output attributable to the general contracting division increased between 1982 and 1985 (see also Tables C15 and C16 - Appendix C). This mirrors the trend in employment in this division, as seen above.

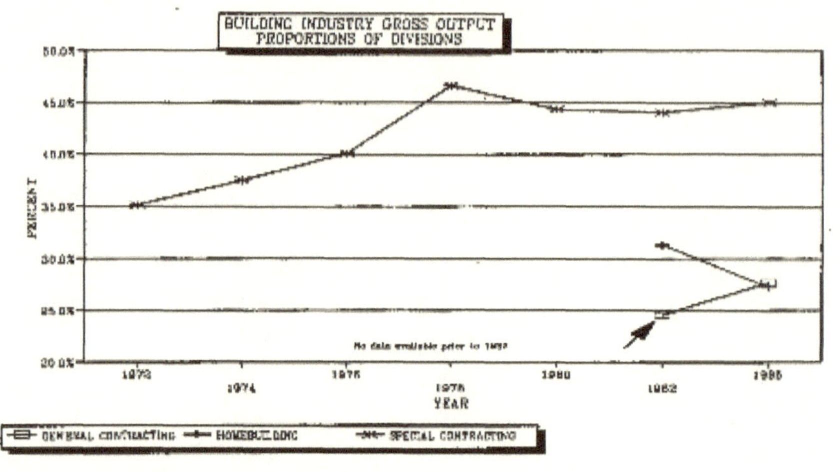

Source : Census of Construction (1985: 1-2)
Figure 3.13 - Building Industry Gross Output - Contribution of Divisions

3.4.3   Gross Domestic Fixed Investment in Buildings

As was noted in section 3.2.4, the most important feature of the construction industry in the economy is contribution to gross domestic fixed investment.

Investment in buildings derives from demand in both the public and private sectors. From Figure 3.14 the trend towards declining public sector investment in both residential and non-residential buildings can clearly be identified (see also Tables C17 and C18 - Appendix C). The decrease is more pronounced in the case of residential buildings - public sector proportion of investment in residential buildings decreased from 34.4% in 1980 to 17.2% in 1989. This is related to the withdrawal of government from the provision of housing (Krafchik, 1990:213).

3.4.4   The Informal Sector

Two types of firms operate in the informal sector - contractors and sub-contractors. As mentioned above, the value of work; wages and salaries; and employment attributable to informal contractors are excluded from official South African statistics. Official statistics do, however, reflect the value of work attributable informal sub-contractors (this is included in the value of the principal contractor's work), although wages and salaries and employment are excluded.

Regarding the informal sector, Krafchik (1990:72-74) observed the following :

67

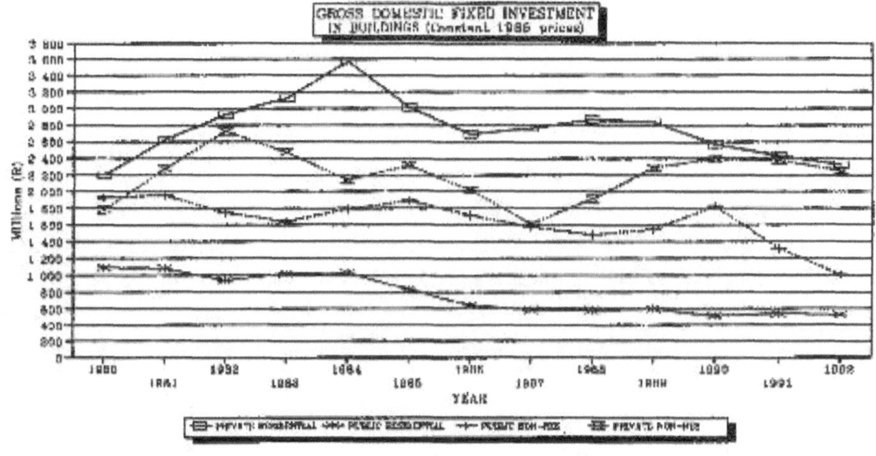

Source : BIFSA (1992: 3-4)

Figure 3.14 - Gross Domestic Fixed Investment in Buildings - Public and Private Sectors

> "...there is no direct way of showing the participation of informal sector contractors in the industry...provides an indirect method to assess the increasing involvement of informal sector sub-contractors. This method involves an analysis of the relationship between value added and gross output in the construction industry.
>
> ...The trend over the period considered [1972-1985] is for value added to comprise a declining proportion of gross output. There are three possible explanations...(i) random errors, which would seem improbable; (ii) capital intensification in the sector, leading to an increase in productivity. This is also unlikely, given the declining real rate of fixed investment in the industry since 1972...(iii) must refer to an increase in sub-contracting to unregistered firms in the industry.
>
> The use of small, unregistered firms would seem most pronounced in the building industry..."

The size of the informal sector cannot be estimated easily. Some insight can, however, be gained from Krafchik's (1990) case study concerning two Western Cape homebuilding projects. Here it was found that 69% of sub-contracting firms were unregistered informal enterprises. Krafchik (1990:111) found that 81% of his sample would have made a loss if they had paid the statutory minimum wages. The incomes of working proprietors in this sample were notably generally higher than would have been applicable if they had been employed as artisans, and substantially higher than the mean for the total informal sector.

The informal sector has been identified as a potential fountain of skills and potential entrepreneurs where even modest assistance can produce significant results (World Bank, 1984:31; Van Staden, 1993). This notion is echoed by Dr. Anton Rupert, vice president of the board of governors of the Urban Foundation who describes the informal contracting community as being "as vital today as the formal construction sector in generating economic wealth, particularly among low-income communities" (UF, 1988). Non-governmental organisations (NGOs) play an important role in the interface with the group targeted for assistance (World Bank, 1984:56). In South Africa, a number of NGOs are, indeed, active in the promotion of small and informal contractors (see Appendix F).

3.4.5   Black Entrepreneurs in the Building Industry

There is no direct manner of showing from official statistics how black contractors interface with the
building industry. As noted in Chapter 2, the majority operate as sub-contractors in the formal sector or as
informal contractors. In addition, it was observed that the majority have had little exposure to the kind of
managerial and supervisory training which is critically important to the success of a SSCE.

In Chapter 2 it was established that many entrepreneurs enter contracting via the *trade route*. These are
individuals who had previously been employees in larger-scale construction firms. It follows that their
positions in these firms would have determined the kind of experience they had obtained. In this regard, it
is interesting to study the data on the distribution of black employees across the occupational groups of:
foreman; artisans and apprentices; production workers and operators; and labourers.

Employees in the occupational groups of 'labourers' and 'production workers and operators' are unlikely
obtain experience in supervision of personnel or management of production. As can be seen from
Figure 3.15, the vast majority of black employees fall into these groups. Artisans and apprentices are likely
to obtain exposure to production management, at least at the level of site-based operations, while foremen

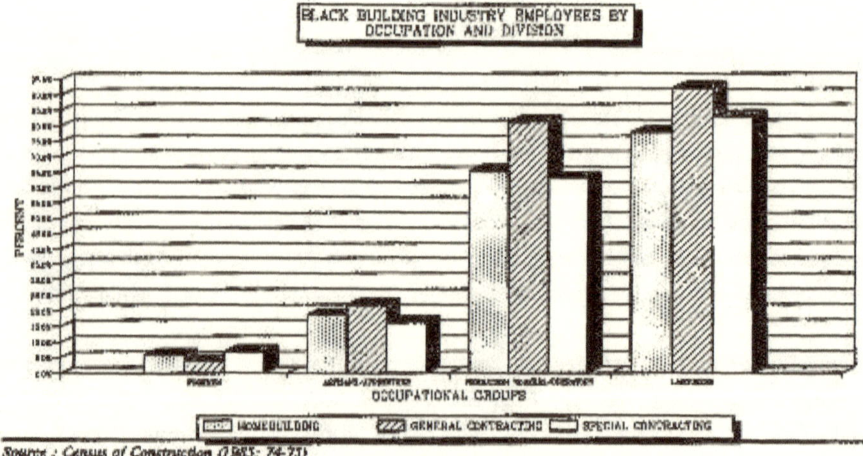

Source : Census of Construction (1985: 74-75)
Figure 3.15 - Distribution of Black Employees Across Occupational Groups

are the only grouping likely to obtain experience in the management of personnel and production
management. Table C19 (see Appendix C) and Figure 3.15 make it abundantly clear that Blacks employee

by construction firms as foremen and artisans and apprentices are in the minority, with proportions remaining fairly constant across the divisions of the building industry (between 4% and 7%). It is clear therefore that black employees do not hold a major stake in managerial and supervisory positions in the building industry.

3.4.6   Training in the Building Industry

Given the nature of building work, it is necessary for firm owners to have a sound knowledge of the building assembly processes. Thus, access to skills training is important to the aspirant construction firm owner. In this regard it is of interest to briefly review training in the building industry. At the outset it must be remembered that only employees of registered firms have access to BIFSA training facilities (whether or not they are employed by MBA member firms) and that most black contracting firms are not registered.

Skills training for the building industry is available through the Building Industries Training Scheme (BITS) which is funded through a levy of 1½% of wages on employees of registered firms. Unemployment training (using BITS instructors and methodology) is available through the Department of Manpower (Merrifield, 1992:75). Numbers of trainees have declined sharply in recent years, particularly in the case of artisans. According to Merrifield (1992:77), BIFSA acknowledges that its training record is "far from satisfactory", with estimates indicating that approximately 80% of formal building industry members have had no training. Indications are that on-the-job training is the most common method of training. The majority of homebuilding firms interviewed by Merrifield (1992:77) indicated that they made little use of BIFSA training facilities, tending rather to train employees with basic unskilled site experience as skilled workers.

BIFSA training requires that workers be released from site to attend the various training modules. Merrifield (1992:78) notes that sub-contractors are the largest group requiring training. Further, given the low margins in this market, it is unlikely that these firms would utilise a training scheme which required them to support their employees for the duration of the training. The volatility of the construction sector, and particularly the homebuilding market, creates a climate in which employees rarely remain employed in any particular firm for long enough to learn a variety of skills (Merrifield, 1992:80). This in turn impacts negatively on the transference of skills from on-the-job trained skilled employees to their operatives and a likely result is that the general standard of skills will diminish progressively over future years. In Merrifield's (1992:80) view, the reformulated BIFSA training scheme "does not provide any answers to this problem".

CHAPTER 4

THE EFFECTIVENESS OF THE INSTITUTIONS SUPPORTING SOUTH AFRICAN SMALL-SCALE CONSTRUCTION ENTERPRISES

The purpose of this chapter is to analyse the existing infrastructure formed by those institutions which specifically strive to educate, train, or in any other way facilitate the development of black SSCEs in South Africa, in order to identify its intrinsic deficiencies. Details of the history, purpose and achievements of these institutions can be found in Appendix F. It is important to clarify at the outset that a considerable number of institutions are involved directly or indirectly with the support of SSCEs. This chapter focuses on a selection of these, where the main criterion for inclusion is the availability of data about the organisation. Failure to mention an institution should not be construed as meaning that its activities or aims are deemed to be inconsequential, but rather that it was impossible to collate sufficient useful data for inclusion in this dissertation.

It is assumed that the average reader will already be well acquainted with most of the institutions described in Appendix F. For these readers, it will be possible to read the body of this chapter without first having studied this appendix. It is, however, recommended that readers less familiar with the institutions do undertake such a study.

For the purposes of this analysis, support institutions have been confined to those that interface directly with SSCEs and are categorised as follows:

(i)   Contractors' Associations (CAs)
(ii)   Profit-making development institutions (PMDIs)
(iii)   Non-profit-making development institutions (NPMDIs)

4.1   THE EFFECTIVENESS OF THE SUPPORT AVAILABLE FROM CONTRACTORS' ASSOCIATIONS

Perhaps the most important observation that should be made here is that the Building Industries Federation of South Africa (BIFSA) and the National Association of Home Builders (NAHB) (hereafter collectively referred to as the *formal industry*) are not active in the *support* of non-member black SSCEs. On the contrary, it is indeed a commonly held perception amongst informal contractors that the formal industry's contracting members, particularly the large-scale white-owned developers and material supply enterprises, are directly responsible for many of their misfortunes. Further, black sub-contractors reportedly see little advantage in joining BIFSA through an affiliated Master Builders Association (MBA) and believe that if they did join, they would "remain marginal" (Merrifield, 1992). Merrifield (1992) goes as far as to say that, his impression is that "despite their protestations to the contrary, the large contractors have no idea, and little interest, where the training of their workforce takes place, as long as the job gets done". This effectively saddles the SSCEs with the role of providing on-the-job-training for their employees.

It seems obvious that the emergence of the black CAs was a response to the formal industry's lack of interest in the informal contracting sector, or of its resistance to losing control over how it should express this interest. A line has thus been drawn between the formal and informal contracting sectors - or more precisely, between the formal industry and the black CAs. It must, however, be pointed out that the formal industry might well be divided internally between liberal and conservative blocs, where the liberal bloc genuinely desires the integration of the formal and informal sectors, and the conservative bloc opposes this. Nevertheless, BIFSA's views on and involvement with the informal sector, or lack thereof, would essentially be controlled by whichever bloc is in the majority at any given time.

An important difference between the formal industry and the black CAs is that in the case of the latter, their membership is essentially made up of contractors (probably with the majority being sub-contractors), whereas the formal industry includes contractors, manufacturers and suppliers. The formal industry obtains its income from this range of members - and serves their interests. The obvious question that this raises is - can it be expected of the formal building industry to contribute towards the development of small informal contractors if such development would ultimately manifest as upward pressure on labour costs, increased competition, *etc.*? One would think not, and this is indeed evidenced by the fact that a wide gulf has existed, and currently exists, between the formal and informal sectors. And it is highly likely that upward pressure would be exerted on labour costs if informal labour-only sub-contractors became members of the formal industry (and if part of the formal industry's resources were therefore directed towards the specialised development and support of these enterprises), for two reasons.

First, as members of the formal industry, small sub-contractors would be under greater pressure to register with Industrial Councils and to pay minimum wages to their employees. This, in turn, would increase their need to recover these labour costs from their employers, the principal contractors. In other words, the exploitive tendencies (Krafchik, 1990; Merrifield, 1992) of principal contractors in their relations with sub-contractors would face increased resistance if the numbers of sub-contractors within the organisation increased significantly. It must be noted, however, that mere membership of BIFSA does not seem to affect whether or not principal contractors will attempt to underpay sub-contractors, since this practise reportedly occurs with member sub-contractors.

Secondly, as the effects of support and development ventures begin to take root, one would expect the improved management capabilities of the small firms to enable them to improve their ability to forecast real labour outputs and costs - and in so doing, reduce losses due to this common weakness. It is also highly likely that an improvement in the expertise and reputation of the informal small contracting sector would essentially represent an increase in competition to the formal small- and medium-scale contracting sector.

One could, therefore, hardly expect the rank and file of the formal sector to be too enthusiastic about the integration into their ranks of the informal sector contractors. A further point which deserves considerati is that the formal sector might well contain within it pockets of resistance to Blacks entering what they perceive to be a white arena - sophisticated quality construction.

It can be clearly seen that there is a strong argument which suggests that the formal sector controls the building industry (SAHAC, 1992:182) and that the informal sector exists precisely because the formal sector wants it to, that is, its existence serves the formal sector's interests. The major black CAs are at present liaising with the formal industry over the issue of unity within the whole industry. Their position are, however, fundamentally unequal in terms of both representativeness (diversity of types of member firms) and regularity of income (amount of recurrent income from members). They are essentially very different kinds of contractors' representative bodies: - one which represents the current owners of the formal industry; and a group of others which collectively represent firms whose informality confines thea to the markets which the first group cannot, or choose not to, dominate. What then is the attraction betw these somewhat unlikely partners? To answer this question, another can be asked - why has the formal industry only now (early 1990s) shown an interest in the informal sector? A tentative answer is that it is only now that significant amounts of foreign funding seem set to be directed at the developing componen of the South African economy. And the targets of this funding, as it affects the building industry, are housing and the development of small- and medium-scale enterprises (with special attention being paid to non-white enterprises). Whether or not the formal sector is prepared to admit to this as a motive, it is cle that its interest in, and involvement with, the informal sector is one of the few strategies which could res in it gaining (albeit indirect) access to this funding.

Thus, the outcome of the liaison between the formal and informal sectors is unlikely to be successful unl it involves some radical restructuring within the organisations themselves. The formal industry has alread attempted to entice black SSCEs into the fold, but as long as it is perceived to represent white interests, only attraction to the informal sector will be that it is a possible catalyst for development or a source of funding. Black resistance to the formal industry could be seen in the failure of BIFSA's in-house FEST programme, which, it seems, had to be cut loose from BIFSA (as Entrepreneurial Development Southern Africa - EDSA) in an effort to legitimise it politically.

Turning now to the effectiveness of the CAs described in Appendix F, it is important to note that there a many of them, and that they all claim to serve the interests of small or black builders. If their aim is the advancement of small, black, or any other kind of contractors it is paradoxical that so many different bo exist, since they can essentially only represent competition to one another (ILO, 1987:125). In so doing, they are, in effect, likely to be working to the detriment of the cause they champion. This is said in the sense that, if it is important for these associations to retain their identities *because of something other tha black- or small-contractor representation*, for example, political affiliation or market control, then by retaining their separate identities, they curtail black- or small-contractor unity. And it is only unity and strength in numbers that will enable black contractors to challenge BIFSA's control of the industry.

It could thus be argued that far more would be achieved if these associations amalgamated, joined the formal industry and, as members, forced BIFSA to take notice of their needs and to put its capital to work towards the satisfaction of these. It could further be argued that the problems between informal contractors and the formal materials supply industry could best be handled as an internal BIFSA problem. Nevertheless, for the present, the retention of separate identities appears to be important. This is understandable if one recognises that there appear to be three different types of black CAs - those that bear the names of large housing developments or of geographical locations, those that are independent and those that are affiliated to national black consumer groupings or chambers of commerce.

The fact that certain CAs are named after large housing developments or areas in which such developments have taken place, suggests that they might be lobby groups whose main purpose was initially to obtain sites for their members in large housing developments (HSA, 1989c). These groups may continue to operate after the completion of such developments, but the reason why they were formed in the first place is an important clue that one of the problems faced by small black contractors is the difficulty of obtaining sites in housing developments. While recognising that those associations affiliated to black consumer groupings or chambers of commerce have been in existence for a relatively short period, their stated objectives have not been accomplished on any significant scale. Their successes have been few, for example :- they have risen to a certain degree of prominence, they have secured access to sites for their members, they have facilitated *ad-hoc* training seminars and to some extent have contributed towards the dissemination of literature tailored to the interests and problems of small black contractors.

The major reasons for the limited success of the black CAs have been fragmentation within the sector and insufficiency of recurrent income. Lack of unity has the effect of diminishing their position relative to the formal industry. And until they can secure a source of recurrent funding (for example, from their membership), they will not be able to fund their strategies. The problem is that, unless the CAs can provide their members with what they need, they will not be prepared to pay for membership. Thus, it is argued that the benefits of belonging to a CA are generally not sufficiently attractive (or obvious) to small contractors. If their successes have been limited, does this mean that black CAs have failed? Possibly, but not necessarily. What it could mean is that all is not well within the CAs. What this might involve can be sensed by asking questions such as: - why do they retain their separate identities and what do they get out of this?; and - who do they represent and do they fulfil their mandates?

Why the need for separate identity? There are many reasons why this could be important, for example: market control; political affiliation; ideological position; ethnic identity; *etc*,. There is no hard evidence to suggest that CA leaders hold their positions in order that they might benefit personally, but suffice to say that this could happen. It would, at least, be a good reason for associations to stay separate. CAs could, therefore, effectively be cartels which control access to particular markets. And cartels would only unite

with other cartels if this was expected to be mutually beneficial (ILO, 1987:121). It could also be that the
associations provide an easily accessible and reasonably prominent political platform, but it is difficult to
establish the extent to which this occurs. Equally possible is that CAs might be split along capitalist-Marx
lines, Xhosa-Zulu lines, etc.. In discussions with a NAFBI leader and a SSCE training facilitator during
July 1993 the author learnt that the African Builders' Association (ABA) and NAFBI had just agreed to jo
forces under the title of the National African Building Contractors and Allied Trades Forum (the NABCA
forum). This is certainly a positive step, but it remains to be seen how representative it will be.

How representative are the existing CAs? It is possible that membership figures claimed by various
associations are exaggerated. But, how many black SSCEs are there? ABA claimed in 1990 that it
represents 95% of black builders (ABC, 1990d). Therefore, if ABA has 3000 members, there must be 3 1
black builders in South Africa. The National African Federated Chamber for the Building Industry
(NAFBI), however, potentially has 2500 members (ABC, 1990b). Does this mean that the bulk of ABA's
members had defected to NAFBI?, or that ABA was mistaken about the total number of contractors in
existence? The truth is probably that no-one really knows the number of black building firms in existenc
But, adding ABA's 3000 to NAFBI's 2500 plus, say, 3000 for all the others, it appears that 8500 black
contractors are probably represented in one way or another.

Can we, however, know whether or not all contractors belong to associations? If only one-half belong, th
there must be 17000 black builders, if only one-quarter do (as suggested by the survey findings in Chapte
5, Table 5.3), then there must be 35000, etc.. Against this speculative background, it can be seen that it
quite possible that the existing CAs are indeed very unrepresentative of black builders. On the other hand
if the figures are correct and there are only about 8500 black contractors in the country, it would then see
excessive for 12 (see Appendix F) unaffiliated institutions to represent them.

Thus, it is argued that the effectiveness of the CAs is largely confined to the fact that they have become
fairly widely known. In other words, it is largely because of the existence of these associations that an
awareness has been created that small black contractors exist in significant numbers, that they have specia
kinds of problems, and that they wish to be free to develop into successful firms capable of operating in
formal markets. In terms of the ILO thinking on this subject (summarised in section 2.6.5), the black CA
are moderately successful in the area of 'external representation', but are unsuccessful in the areas of
'internal services' and 'standards setting'.

4.2 THE EFFECTIVENESS OF THE SUPPORT OFFERED BY PROFIT-MAKING DEVELOPMENT INSTITUTIONS

The Kwazulu Finance and Investment Corporation (KFC) is something of a special case, since it is
essentially non-profit-making in its operating style. Its subsidiary, Kwazulu Training Trust (KTT), plays

far more direct role in the training and support of SSCEs, but since it is a non-profit-making institution, it is discussed in the next section. The impression might be gained from the following critique that KFC is being unfairly likened to other PMDIs. This is not intended, and where a point does not apply to KFC, or for that matter, any other institution, it is hoped that the reader will recognise this.

The Perm subsidiary, HODECO, is also a special case because it no longer exists. However, in its time it was one of the relatively few programmes of its type. Its inclusion is necessary to facilitate the discussion below, since what is perhaps most important about HODECO, is that it failed. It is thus included below, and referred to as if it still existed, notwithstanding its liquidation. There are also a number of private sector development companies whose involvement in low-income housing production, and therefore in the world of black SSCEs has been important. To include these firms in the following discussion, or the materials suppliers that have provided special assistance to black SSCEs, and so on, would exceed the intended scope of this dissertation.

The PMDIs are involved with small contractors for a variety of stated and unstated reasons: - profit; market expansion; entrepreneurial development; regional development; *etc.*. But since the basis of their definition as 'profit-making' is that they pay dividends to shareholders, it is argued that the need to generate profits and therefore dividends fundamentally affects how they operate and, indeed, whether or not they operate. While their greatest success has been the partial alleviation of the problems of SSCEs, their greatest failures have been their inaccessibility and non-sustainability. Although as organisations they differ widely, they are not separated in the following critique, but rather, their generic similarities, strengths and weaknesses and collective effectiveness are highlighted.

The Small Business Development Corporation (SBDC) and KFC have something in common - the fact that their activities are not confined to the development of SSCEs. This makes it less important that their small contractor programmes succeed than it is for the housing-oriented SA Housing Trust (SAHT) and HODECO. Put another way, the success of the organisation as a whole is, in the case of these two institutions, not dependent on the success of their SSCE programmes. Nor, however, is it essential for the SAHT and HODECO to have successful small contractor programmes, since they, respectively, provide houses, and finance houses. In both cases, involvement in the training of SSCEs is optional. Since they possess money, but lack two internal resources - builders and home-buyers - it is clearly in their interest to assist builders and buyers in getting together - and if improving the capabilities of builders can help this, then SSCE development begins to look like a sensible investment.

The SBDC's interest in small builders is thus similar to HODECO's in that both are essentially financiers. Their motives may truly be to develop SSCEs, but in addition to whatever else they accomplish, these two companies facilitate the involvement of SSCEs in what (for the SSCEs) is usually a prohibitively difficult process, and are rewarded in the form of interest on a loan - to the builder in one case, and to the home-buyer in the other. In fact, HODECO never actually achieved this reward and always ran at a loss - which

was underwritten by the Perm. Since the overriding motive of the financiers is profit, they must protect
by not exposing themselves to situations that look too risky. And, to them, all situations involving black
home-buyers or black SSCEs, are potentially very risky. The unfortunate outcome of this tendency to av
risk is that they only reach a very small number of the SSCEs that really need their help. It would thus
appear that the common claim of SSCEs that lack of access to finance is their greatest problem, is well-
founded.

The main evidence of risk avoidance can be seen in the criteria that determine acceptance onto small buil
programmes. The qualifying criteria set by SBDC, HODECO and SAHT are so high that the vast major
of small contractors would be refused entry on application (this has been confirmed by Merrifield (1992)
The black SSCEs interviewed by Calcopietro (1992), however, seemed to be unaware of the exact
requirements of the SBDC. This suggests that communication about its small contractor programme need
to be improved. It cannot be said that the SBDC has been successful in its attempts to provide services to
black SSCEs. For example, in 1992 there were no black contractor clients of the Western Cape SBDC,
while in Natal there were only a few. The manager of the Western Cape branch reported that the SBDC
regarded this as a problem and was attempting to improve matters by addressing the issue in EDSA's Sm
Contractors' Action Forum (SCAF) (Calcopietro, 1992). The SBDC personnel in Durban who administe
the small builders' bridging fund reported in informal interviews with the author that their experiences w
small contractors had been so negative that the programme had virtually been suspended. At the time of
writing, new moves were afoot to revive the scheme.

The consequences of risk avoidance have not gone unnoticed by black SSCEs, who have criticised SAHT
for using large established developers in its early developments. As was noted in Appendix F (F.B.2(e)),
only 9% of the work on SAHT developments to date has been executed by SSCEs. HODECO's avoidan
of risk is built in - it can control the number of home-buyers it creates and it can commission SSCEs to
build homes for them. Its final collapse was precipitated by its loss of control over the creation of home-
buyers, or put another way, its markets became so affected by township violence that it had to stop
approving home loans. In so doing, it made its scheme unworkable. Merrifield (1992) has also noted tha
high degree of conservatism and inflexibility accompanies risk avoidance and cites as an example the
SBDC's reluctance to reward long-standing clients (with good repayment records) by tailoring its credit
facilities to their credit needs.

Similar to risk avoidance is the avoidance of dependency. SAHT's main purpose is to provide housing, y
it considers it necessary that SSCEs involved in its projects ultimately progress to the level of developers
their own right. This suggests that, as well as producing houses, the SAHT produces a by-product -
developers. This may perhaps be too liberal an interpretation of Duncan's (1992b) statement that "the ro
to self reliance...must be continuous, i.e. from Small Builder to Mini Developer to Developing Agent".
the main thrust of SAHT's involvement in SSCE development is that there should be a gradual decrease
the dependency of the contractor on SAHT. In other words, throughput is important.

The average SSCE, however, probably has more interest in SAHT as a client that creates a demand for houses in the development market, than as a training experience (Wells, 1986:111). This is not to say that the training experience is not regarded as beneficial, but rather that the contractors who have been involved in SAHT projects have been going about their business - contracting. SAHT, however, apart from being a client, assumes the additional role of facilitating on-the-job-training, thus providing far more in the way of support than the average client in the housing market.

We can thus see that this overriding ideological influence in the SAHT mission makes it a rather atypical client in the housing market - from the perspective of a small contractor. SAHT's desire to produce mini-developers as a by-product of housing, conflicts with the needs of many SSCEs that, for example, wish to stay small, but busy. Indeed, one of Merrifield's (1992) interviewees raised the question of the popular obsession with transforming black contractors into independent general contractors, pointing out that more could perhaps be achieved if their capacity as sub-contractors was improved. In other words, some SSCEs have no objection to operating as sub-contractors and do not want to be pushed into the role of principal contractors or developers. For these firms, the SAHT approach is difficult to understand.

Exploring this idea further, it can be argued that, once SAHT has created numerous mini-developers, it will have created competition for white developers - whose dominance of the market was part of what prompted the formation of organisations like SAHT in the first place. This makes it difficult for SSCEs to understand SAHT's collaboration with established developers.

There is evidence, however, that SAHT and HODECO have not been successful in the creation of mini-developers. In this regard, Merrifield's (1992) observations (already noted in section 2.6.4) about such programmes being too controlled and SSCEs relating to them as if they were private sector developers will be recalled. Merrifield's (1992) caveat was that such programmes could create a 'culture of dependency'. This tends to be supported by the fact that, by 1992, SAHT had produced a total of 112 small builders, but only 10 mini-developers.

HODECO is also intrinsically interested in throughput. Again, this is not to say that HODECO does not solve one of the biggest problems for SSCEs and home-buyers alike, that is, the access to finance issue. It does, but only for proven contractors and hand-picked home-buyers. Once this marriage has been arranged, HODECO moves on to the next. As was described above for SAHT, HODECO is regarded by the SSCE as a client in the housing market. This is evidenced by its provision of serviced sites and buyers with guaranteed loans - it effectively puts orders for houses into the market place of the SSCE. However, because it provides a serviced site and a home-buyer, HODECO, like SAHT, does more than the usual commercial client does. This might be an attempt to penetrate black residential markets in a legitimate manner, but then it might also be intended as a venture which is worth doing because it benefits all parties.

To summarise, the main factors which negatively affect the success of the PMDIs described above are (i)

that their entry criteria for SSCEs are prohibitively high, denying access to support to the majority of th
who most need it (risk avoidance), (ii) that their missions are not understood by SSCEs - who regard so
of them as clients or sources of work - hence a basic resistance by SSCEs to the notion of throughput
(dependence avoidance), and (iii) that it may not be vital to the success of the institution as a whole tha
SSCE programme succeeds - which results in little effort being made to ensure its success.

There are, however, also positive aspects to the involvement of PMDIs in SSCE development. At the v
least, each programme that has in any way provided work opportunities for SSCEs and their employees
and has therefore also provided opportunities for on-the-job-training - has contributed towards the
development of SSCEs and, in some cases, to the reduction of the housing shortage.

This is especially true in the case of KFC, SAHT and HODECO, who actually produce houses, service
sites, or provide finance to home-buyers. The specific stimulation of the process of getting houses built
very important contribution towards the solution of the training crisis. Too many informal contractors
already run businesses and would never be able to attend lengthy training programmes for the existing
training infrastructure to be capable of accommodating their special needs. Exposure to the contracting
process, and the chance to build up experience (the lack of which have been described by Merrifield (19
as the root of many of the problems experienced by black SSCEs), represent the only real access to trai
or support for such contractors, *provided they can get on to the programmes*. The promotion of self-reli
is another positive contribution from the PMDIs. This is especially helpful where self-reliance is not tak
to mean the attainment of developer status, but rather means the ability to operate at any level without
needing support.

In summary, the positive contributions of PMDIs to SSCE development are mainly, (i) that they provid
access to work opportunities (and therefore experience) and (ii) that they amplify the existing training
infrastructure by providing access to on-the-job-training/counselling.

What is more important than the specific successes and failures of these organisations is that they perfor
function which caters well for the needs of SSCEs of a certain (high) calibre. They do not target the wh
range of SSCEs, since other organisations could play that role. The value of their contribution must
therefore be seen for what it is, rather than for what SSCEs or anyone else would prefer it to be.

4.3 THE EFFECTIVENESS OF THE SUPPORT OFFERED BY NON-PROFIT-MAKING DEVELOPMENT INSTITUTIONS

The Shelter and Urban Development Support Programme (SUDS) described in Appendix F (F.C.7) lool
like a promising step in the right direction. But, since it is not yet in operation, it will be excluded from
following analysis. It must, however, be noted that its South African Black Contractors Assistance

Programme (SABCAP) initiative targets black SSCEs "who have growth potential and resources to operate a formal contract", along with its approved plans, cost estimates, *etc.* (CUSSP, 1993). These criteria could effectively make it inaccessible to the majority of informal contractors.

The NPMDIs are involved with small contractors for similar reasons to the PMDIs, namely: entrepreneurial development; skills development; and regional development; - but with the obvious exceptions of profit and market expansion. At least, they are not into market profit maximisation and market expansion in the usual sense. They certainly do compete for the same market, that is, SSCEs and those organisations that grant resources for SSCE development. In addition, many do attempt to make profit, except it is called surplus. But, there is a more indirect 'profit' that they make - overheads. Many SSCEs are quick to notice that the personnel of some of these 'non-profit-making' institutions get good salaries, drive expensive company cars, operate from plush offices, *etc.*. However, their motives must essentially be assumed to be development-oriented, with their special focus being self-reliance and sustainability. They tend to have a greater interface with the informal sector than the PMDIs, and are considerably more training- and community-oriented. Their greatest successes have been the facilitation of skills endowment and the provision of work opportunities, but like the PMDIs, their greatest failure has been their relative inaccessibility and potential to be non-sustainable.

There is some evidence of risk avoidance in the activities of the NPMDIs. For example, the Urban Foundation's (UF) Contractor Development Agency (CDA) programme only caters for "A-Grade" contractors and is relatively expensive to enter. Thus, by this and various other means, some of the NPMDIs restrict their intakes to enhance their chances of success. However, there is also an element of what could be called 'risk invitation' - as is seen in the employment of community-based and informal contractors by The Newco Group (Newco) and Tholiwe Homes (TH).

The risk that accompanies the commitment to work with community-based and informal contractors, is diminished by the provision of training and the exertion of some influence or control over the SSCEs. The facilitation of access to finance (by CDA, EDSA, KTT, and TH) is also potentially risky. But, since this always appears to be the biggest need of SSCEs, the institutions which purport to assist them, have little alternative but to address it. A major reason why many NPMDIs can afford to invite risk, is that they can transfer this risk to their benefactors. This can be seen in the relationship between: the former Housing Utility Companies (HUCs) and the UF; and EDSA and the Development Bank of Southern Africa (DBSA)/ First National Bank (FNB)/International Business Machines (IBM), *etc.*. It can, however, be seen that risk transference will only occur to a limited extent before the benefactors revise their strategies. For example, the formation of Newco and the resultant shift in HUC emphasis to site-and-service and squatter upgrading schemes, was necessary to avoid the UF incurring ongoing losses. Thus, the Newco strategy could represent a shift towards a type of involvement that excludes having to facilitate the construction of conventional buildings (and therefore avoids having to deal with conventional builders with their concomitant financial and other needs).

As was seen with certain of the PMDIs, success in SSCE involvement is not necessarily critical to the success of the institution as a whole. This is the case for the CDA programme and the KTT and Palabora Foundation (PF) training programmes. It was noted above that certain of the PMDIs tended to promote throughput, and that this was resisted by SSCEs who regarded them as clients in a housing market place. Similar tendencies are evident among the NPMDIs (for example, CDA, EDSA, KTT, and PF), but since they perform more of a training role than a financing role, they can legitimately promote throughput and self-reliance. They, nevertheless, tend to be more accepting of the right of small firms to stay small, and generally do not mirror the tendency of the PMDIs to want to transform every SSCE into a developer.

It could be seen as a shortcoming that certain NPMDIs create an unrealistic market environment for SSCE. This is seen in the CDA programme where the contractor is *entitled by a contract with CDA* to a steady inflow of work. TH's provision of materials and its attempt to eliminate the sub-contractor also make it a atypical client. The success of these strategies would ultimately be judged in terms of the ability of the SSCEs they produce to retain the skills they have acquired, once they move out into the real (unsupported) markets. If SSCEs fail to retain these skills, the programmes would effectively have created dependants, rather than skilled entrepreneurs.

Another factor which could retard the success of certain NPMDIs is their direct, or indirect, connection with BIFSA. As noted earlier, there is resistance on the part of black SSCEs to BIFSA affiliation. EDSA has only recently moved from the free office space provided to it by BIFSA (on BIFSA's premises) (EDSA 1992) and indeed, to some extent, its directorship has, over the years, reflected its BIFSA origins. The BIFSA-connection can also be seen in the training programmes of KTT, PF, and Bloemfontein Building Industry Development Foundation (BBIDF), which either are modelled on BIFSA's 'competency based modular training' (CBMT) system, or require collaboration with BIFSA-affiliated MBAs. So, the BIFSA-connection is politically disadvantageous, but there are further considerations such as: the level at which BIFSA-style courses are pitched; and the duration of these courses. Although they serve a very valuable function by providing training to new or unemployed members of the black SSCE fraternity, the duration the courses is too long for operational SSCEs to attend. As was noted above, these firms need access to on the-job training. The severe decline in BIFSA's artisan output in recent years (Merrifield, 1992:67-68) would seem to suggest that the market simply does not require skills levels to be of the standard of the BIFSA CBMT programmes. There is a large presence of informal SSCEs in the contracting market and their training needs must also be addressed - even if this means dropping the standards maintained by BIFSA.

A common failure of virtually all of the NPMDIs, is that, in a number of ways, they are inaccessible. Inaccessibility takes various forms including: limited geographical coverage (most have only made any impact in just one region or urban area); exclusive curricula; expense of entry; *etc.*. To SSCEs, it appears that a number of institutions are trying to help them, but in effect, they hardly seem to scratch the surface. Some fall short of providing what is really required. For example, Merrifield (1992:71) has noted that "th 'entrepreneurial' training offered by EDSA is not specifically designed to train sub-contractors as supervisors, and, while [it] may assist them in their relations with formal contractors, their training, by design, does not address the needs of big site operations". Whether or not exclusivity is consciously built into the courses, it is possible that many SSCEs find the content of curricula inappropriate or deficient.

Those institutions that have solid financial backing (for example, Newco, EDSA, KTT, PF) are relatively less vulnerable to the dangers of rapid expansion, in terms of the extension of geographical coverage or the adaptation/extension of curriculum content, than those that are still trying to arrange recurrent financing (for example, TH). Unless TH manages to secure recurrent funding, its activities will have to be confined to self-funding projects. This is not to say that non-self-funding projects should be endorsed, but it does mean that TH would not be able to do anything at all unless it had a viable housing project under development. In other words, expansion would be governed by the number of viable projects available.

There is also the problem that certain areas (for example, the Cape Province) have a substantially less developed infrastructure of support institutions for small black SSCEs because of their relative remoteness from areas in which the various statutory development institutions have been operating. In these areas the problem of inaccessibility is aggravated. EDSA's franchising idea could be a viable and important method of improving the accessibility problems of all institutions with similar products.

To summarise, the main factors which negatively affect the success of the NPMDIs are (i) that they risk creating dependencies, (ii) that the training provided by some is pitched at too high a level for established informal SSCEs and, (iii) that they are relatively inaccessible.

The positive aspects to the involvement of NPMDIs in SSCE development mainly centre around their success in reaching informal and community SSCEs, albeit to a limited extent. CDA is a notable exception since it aims to breed competition for white developers. But, as was the case with the PMDIs, some provide work opportunities and on-the-job training for SSCEs and their employees, and contribute towards a reduction in the housing shortage.

Like the NPMDIs (KFC, SAHT and HODECO), Newco and TH actually produce houses, service sites, or provide finance to home-buyers. As was noted above, this helps alleviate the shortage of on-the-job training opportunities. This is most important for the established informal SSCEs and the specific efforts of Newco, TH and EDSA to involve or train community-based and informal contractors is a positive trend. Another positive aspect of this is the production of contract documentation, *etc.* that is appropriate to the needs of these less sophisticated contractors. Such efforts are reported by Merrifield (1992) to have the support of black SSCEs.

Further positive contributions from the NPMDIs include the contributions to formal housing research by the former Residential Development Division (RDD) of the UF and the less formal research/dialogue-promotion facility which exists in the SCAF. However, Merrifield (1992) has noted that the SCAF is exclusively concerned with the problems of informal contractors and is unwilling to promote the cause of SSCEs working within the formal sector. This has the undesirable effect of excluding black sub-contractors who work almost exclusively with formal sector principal contractors - and the forum "seems constitutionally unwilling" to alter this position (Merrifield, 1992).

The main positive contributions of the NPMDIs to SSCE development are the same as was found for the PMDIs and their additional contributions can be summarised as follows: (i) that certain of them interface with community-based and informal SSCEs; (ii) that they are generally training-oriented and; (iii) that th undertake and disseminate research into the problems of SSCEs, housing, *etc.* and promote dialogue with the informal contracting sector.

It was suggested above that the PMDIs cater reasonably well for the needs of SSCEs of a reasonably high calibre and that their choice not to interface with certain sections of the range of SSCEs was affected by possibility that other organisations could play that role. It can be seen that this role has been adopted by certain of the NPMDIs, but their effectiveness is seriously limited by their relative inaccessibility and dependency on the state of the homebuilding market.

4.4   SUMMARY

The potential of the PMDIs to be successful supporters or developers of SSCEs is circumscribed by their profit motive. This requires that they avoid risk, which essentially means that they have to avoid the majority of the group that they claim to support. This avoidance of risk takes the form of setting entry criteria for support programmes at such a high level, that only those who are already successful (in relati terms) are admitted. In addition, risk is reduced by broadening the scope of their operations beyond pure SSCE support.

For the PMDIs, throughput of SSCEs on their programmes is important. This suggests that the more SSC they can assist, the better they look - and the better they are placed to attract more work on the basis tha they assist SSCEs. The necessity for throughput is also the root of the apparent obsession with transform all black contractors into independent general contractors and the notion that sub-contracting is bad becau it generally occurs under exploitive circumstances. The PMDIs have failed to uplift the black SSCEs and they risk retarding black contractor development by engendering dependency.

The positive contribution of PMDIs to SSCE development has essentially been the by-products of their m function of profit acquisition, that is, the provision of work opportunities for SSCEs and their employees and the creation of opportunities for on-the-job-training. However, these by-products effectively only rea a minority of SSCEs.

The NPMDIs' motives are essentially development-oriented and they tend to have a greater interface with the informal sector than the PMDIs. In addition, they tend to be far more training- and community-oriented. They, like the PMDIs, also avoid risk, but some of them actually invite risk. In some cases the avoidance of risk is overt (high entry criteria, expense), while in other cases it is more subtle (changing markets, for example, from conventional housing to site and service). Those that invite risk by getting involved where they are most needed, for example in housing projects where they support communities and the informal contracting sector, incur a training liability for very different reasons than the PMDIs. In this case they need to train to ensure the success of projects, while the PMDIs do not really *need* to train at all, they just choose to. Another form of risk invitation is the advancement of finance to community-based or informal contractors.

Like the PMDIs, some NPMDIs are not reliant on the success of their SSCE programmes for their overall success. Further, because they are more training-oriented, the NPMDIs tend to attach importance to throughput, but they reach a greater range of SSCE types than do the PMDIs. Since the NPMDIs that involve SSCEs in actual construction work are effectively the clients to the SSCEs, they are at risk of creating SSCEs that can only operate under the extremely supportive and relatively comfortable conditions that they create.

There appears to be a fairly wide endorsement of BIFSA's CBMT methods - sometimes tacit, sometimes overt. Two issues flow from this. First, the CBMT has built into it an element of class - who defines 'competency'? Part of BIFSA's problem with the informal sector is that it would seem to have a different idea of what acceptable quality is, and therefore, what competency is. BIFSA's trained artisan output has declined steadily in recent years, which can be taken to mean that there is effectively little demand in the building industry for that level of skill. Without going any deeper into this important issue, it can be seen that any connection with BIFSA in the training arena could attract a negative reaction from the informal sector - on the grounds that it is white, bourgeois, elitist, *etc.*.

In the case of the PMDIs, access to the programmes was mainly determined on the basis of proven capability (the "A-Grade" contractor syndrome). Although there is an element of this in the NPMDIs, their inaccessibility is governed by their geographical distribution and their financial position (or access to recurrent income). Thus, although for different reasons, they, like the PMDIs, are relatively inaccessible.

The most positive contribution of the NPMDIs' involvement in SSCE development is that they reach informal and community-based SSCEs, providing work opportunities and on-the-job training for SSCEs and their employees. The facilitatory role that certain NPMDIs have played through their contributions to research and the dissemination thereof, and the promotion of dialogue between development institutions, has also been very positive.

A conclusion on the effectiveness of the institutions discussed in this chapter can be found in Chapter 7.

THE SURVEY

CHAPTER 5

ANALYSIS OF THE SURVEY DATA

The purpose of this chapter is to present and analyse the survey data. The chapter is divided into twelve sections (5.1 to 5.12) which correspond with the twelve modules of the questionnaire (A to L).

5.1. SURVEY SAMPLE

In the background to the definition of the problem (section 1.2.1), it was noted that a limitation of Krafchik's (1990) work was that it concerned only sub-contractors. A stated intention of the present study was to include a variety-of-types of SSCEs in the survey sample. As will be seen in Table 5.1, this was accomplished. A total of 96 interviews were conducted - 50 in Durban and 46 in Johannesburg.

TYPE OF CONTRACTOR	- main activity or trade	NUMBER SURVEYED	% OF SURVEY	TOTAL % OF SURVEY BY TYPE
General contractors	- mainly new work	17	17.7	
	- mainly alterations and additions	12	12.5	30.2
Specialist sub-contractors	- electrical	6	6.3	
	- plumbing	8	8.3	14.6
Labour-and-material sub-contractors	- combination of trades	30	31.3	31.3
Labour-only sub-contractors	- carpentry	4	4.2	
	- plastering	2	2.1	
	- painting	1	1.0	
	- combination of trades	7	7.3	
	- brickwork	9	9.4	24.0
TOTAL		96	100.0	100.0

Table 5.1 - Types of Contracting Firms Surveyed

In section 2.4.2, definitions were given of different types of contractors. These definitions apply to this table. From Table 5.1 it can be seen that the survey sample consisted of different kinds of SSCEs. To achieve this range, it was necessary to forfeit randomness. Had a random survey been conducted it is likely that the sample would have been radically skewed towards one or another type of contractor. For example, three quarters of the sample might have turned out to be bricklaying sub-contractors. However, within each sub-group of type of contractor, respondents were selected randomly.

In section 2.4.4 it was determined that the 'alterations and additions' market was one of the few to have

grown in recent years, and that it is likely that SSCEs have been, and currently are, active in it (ILO,
1987:28; Krafchik, 1991:3; HSA, 1989b). For this reason it was ensured that the general contractors, wh
in total accounted for 30.2% of the survey, included those whose main activity was new work (17.7%), a
those who mainly undertook alterations and additions work (12.5%). The fact that contractors of the latte
description could be found tends to support the literature.

The 'specialist sub-contractors' in the survey sample consisted of only electricians and plumbers who
usually operate on a labour-and-material basis. This sub-group accounted for 14.6% of the total survey
sample.

The 'labour-and-material sub-contractors' made up 31.3% of the survey sample. These sub-contractors a
difficult to define. The only thing that prevents them from being defined as general contractors is that the
only undertake part of the whole job. Nevertheless, they are probably closer to being general contractors
than sub-contractors.

'Labour-only sub-contractors' accounted for 24.0% of the survey. The distribution of this percentage acr
the trades can be seen in Table 1.5. The weighting of the distribution was loosely contrived to reflect
differing levels of complexity of the work done by the various categories of sub-contractors.

Before proceeding with the presentation and analysis of the survey data, some explanation is necessary
regarding the abbreviations and referencing used in this chapter.

In the tables, the following abbreviations are used to denote the various types of firms surveyed:

> GENERAL = The group of 29 general contractors
>
> SPECIAL = The group of 14 plumbing and electrical sub-contractors
>
> LAB SUB = The group of 16 labour-only sub-contractors
>
> L&M SUB = The group of 30 labour-and-material sub-contractors
>
> ALL = The total survey sample of 96 contractors (unless otherwise indicated, all tables
> reflect data for ALL types of contractors)

Appendix H is a print-out of the survey data and on pages 1 and 2 it will be seen that the data consist of
380 variables. These can be traced directly to the questionnaire, since the variable names correspond to t
question numbers. In some cases it was considered unnecessary to include data in tabular form in the text
this chapter. Instead, reference is made to Appendix H in the following manner:

> (pg.33/AppH;Q-D_1)

This reference means:- refer to page 33 of Appendix H and see question D_1 in the questionnaire for the original question. In the case of open-ended questions, the actual responses of interviewees are recorded in Appendixes M1 and M2. In Appendix H, numbers are used to denote these verbatim responses. These can be interpreted by referring to Appendixes M1 and M2. For example, Question J_4_1 was an open-ended question. It attracted three different responses, which are represented on page 248 of Appendix H, as the values '1', '2' and '3'. The corresponding responses for these values can be found in Appendix M1 under the heading "Labels for variable J_4_1".

Throughout Appendix H, the following coding system was used:

'0' = Zero;   '-1' = Not applicable;   '-2' = Don't know;   '-3' = No response or missing

Values greater than zero represent what they are defined as in the questionnaire (for example, 1 = YES). In the absence of such a definition, they represent real numbers, unless they apply to open-ended questions, in which case they represent label codes.

All tables in this section were compiled from the survey data. In general, the sample size in each table equals the total sample of 96. However, certain questions allowed multiple responses, resulting in sample sizes which are apparently greater than the number of respondents. These occurrences are highlighted in the explanatory notes to the tables concerned. In addition, certain questions followed on from previous questions, providing only the responses of those who, for example, answered in the affirmative. In the case of these tables sample sizes are lower than 96 and correspond with totals which can be identified in a previous table. In certain cases these smaller sub-samples were also required to answer multiple response questions and sub-sample sizes may consequently exceed the apparent number of respondents.

5.2.   GENERAL

In this section the general attributes of the firms are described. These include: age of firm; whether or not the firm or its owner belongs to a contractors' association; whether or not the owner had previously owned a business; and details regarding ownership and partnerships.

5.2.1   Age of Firm

Table 5.2 shows firm ages for the different types of contracting firms. These firm ages were calculated by finding the difference between the date of interview and the date that the firm was established.

Count Row % Col % Total %	LESS THAN 2 YRS OLD	2 TO 5 YRS OLD	6 TO 10 YRS OLD	MORE THAN 10 YRS OLD	Row Total %
GENERAL	1 3.4 50.0 1.0	15 51.7 29.4 15.6	6 20.7 22.2 6.3	7 24.1 43.8 7.3	29 30.2
SPECIAL	0 0.0 0.0 0.0	8 57.1 15.7 8.3	4 28.6 39.1 4.2	2 14.3 12.5 2.1	14 14.6
LAB SUB	1 4.3 50.0 1.0	9 39.1 17.6 9.4	9 39.1 33.3 9.4	4 17.4 25.0 4.2	23 24.0
L&M SUB	0 0.0 0.0 0.0	19 63.3 37.3 19.8	8 26.7 29.6 8.3	3 10.0 18.8 3.1	30 31.3
Column Total %	2 2.1	61 57.1	27 28.1	16 16.7	96 100.0

Table 5.2 - Age of Firm

The majority of firms (53.1%) were found to be between 2 and 5 years old. This is also the case for each of the categories of type of firm. An important inference can be drawn from these findings, namely that these firms are the survivors of the building recession. The majority have been in existence, and operating in a depressed industry, for long enough for it to be assumed that only the resilient would have survived. This should be borne in mind when interpreting the rest of the survey results. Given that the barriers to entry into the occupation of building contracting are low, one would expect a survey conducted during a building boom to reflect a higher proportion of firms less than 2 years old, than was the case in this survey.

5.2.2 Membership of Contractors' Associations

In Tables 5.3 and 5.4 details are given of membership of CAs. Table 5.3 shows the extent to which the firms reported belonging to CAs, while Table 5.4 identifies these. A significant finding is that ABA was far the most commonly cited association. This does, however, not necessarily mean that ABA is the most popular CA, since the survey size was relatively small. Notably, NAFBI was not mentioned at all.

The majority of the firms (75%) did not belong to a CA. This was distinctly more pronounced in the case of labour-only sub-contractors, where only 13% of that group were found to belong to an association. Notably, the specialist sub-contracting group displayed a considerably greater tendency to join CAs.

This finding was expected, since it was noted in section 2.6.5 that SSCEs generally appear to avoid joining CAs (ILO, 1987:121).This means that the majority of contractors operate without the internal services which could, and should, be offered by a CA. Further, this high degree of non-membership means that CAs are unable to perform a standards setting function (ILO, 1987). There thus appears to be little or no peer group accountability among SSCEs. The implications of this for the general reputation of SSCEs has already been highlighted.

Count Row % Col % Total %	NOT A MEMBER OF AN ASSOCIATION	MEMBER OF 1 ASSOCIATION	MEMBER OF 2 ASSOCIATIONS	Row Total %
GENERAL	23 79.3 31.9 24.0	6 20.7 27.3 6.3	0 0.0 0.0 0.0	29 30.2
SPECIAL	8 57.1 11.1 8.3	5 35.7 22.7 5.2	1 7.1 50.0 1.0	14 14.6
LAB SUB	20 87.0 27.8 20.8	3 13.0 13.6 3.1	0 0.0 0.0 0.0	23 24.0
L&M SUB	21 70.0 29.2 21.9	8 26.7 36.4 8.3	1 3.3 50.0 1.0	30 31.3
Column Total %	72 75.0	22 22.9	2 2.1	96 100.0

Refer to question B_2 of the questionnaire for the original question

Table 5.3 - Membership of Contractors' Associations

Count Row %	ABA	BRC	BIFSA	MBA	ECA	SABA	SBF	SBA	BBA	Row Total %
ALL	14 53.8	1 3.8	2 7.7	1 3.8	2 7.7	2 7.7	1 3.8	2 7.7	1 3.8	26 100.0

Refer to question B_2 of the questionnaire for the original question

ABA = African Builders' association
SABA = South African Builders' association
SBA = Soweto Building Association
BIFSA = Building Industries Federation of South Africa
ECA = Electrical Contractors' Association (a BIFSA affiliate)

BRC = Builders and Renovators Club
SBF = Soweto Builders Forum
BBA = Black Builders' association
MBA = Master Builders' association (a BIFSA affiliate)

Table 5.4 - Contractors' Associations - Distribution of Membership

It was noted in section 4.1 that BIFSA does not at present actively support black SSCEs by means of any specific development programme. Consequently, it was argued that there was a high degree of resistance to BIFSA from the black SSCEs. This is confirmed by the finding in Table 5.4, where it is seen that only 5 of the 96 firms (5.2%) were members of BIFSA or one of its affiliates. It was also argued in section 4.1 that

the existing CAs might be unpopular because they have failed to deliver what SSCEs need. This must,
however be seen in the context of the major CAs being fairly young organisations (ABA started in 1988
NAFBI in 1991). It remains to be seen whether or not these associations will, in time, succeed in meeti
their objectives (see Appendix F).

5.2.3   Previous Ownership of Business

As noted in section 2.4.1, there are relatively few barriers to entry into the occupation of building
contracting compared with other self-employed occupations. This is confirmed in the finding that 90% c
firm owners had never before started a business (pg.22/AppH;Q-B_3).

The importance of this result lies in the fact that it suggests that the majority of firm owners obtained th
practical business management experience while operating their current businesses. This is supported by
finding that only 5.2% of the interviewees had been employed in managerial capacities in the building
industry (see Table 5.9). Further, the response to question D_3 indicated that 57.3% of the respondents
never received any formal business management training prior to starting their firms (see section 5.4.3).
Besides these two results, it must also be noted that the demand for business management training far
exceeded the demand for any other type of training, as can be seen in Table 5.12. But, most importantl
the finding confirms the existence of a need for on-the-job management training.

5.2.4   Ownership of Firms

Table 5.5 contains details concerning the ownership of firms. The majority (76%) of the respondents we
the sole owners of their firms. This was also the case for each sub-group of type of contractor.

In section 2.4.4, this was noted as a general trend and this finding was therefore expected. The finding
'general contracting' and 'labour-and-material sub-contracting' firms were more inclined (than the other
groups) to be solely owned, was, however, unexpected. The complexity of the management function wi
a firm is determined by the degree of responsibility or control it is required to assume in terms of the
contract with its employer. Thus, the management of a labour-only sub-contracting firm is less demandi
than the management of a general contracting firm. It was expected that a positive correlation would exi
between the number of partners in a firm and the complexity of its management tasks.

Of the 23 firm owners who indicated that they had business partners, 13 (56.5%) had 1 partner, 8 (34.8
had 2 partners and 2 (8.7%) had 3 partners (pg.24/AppH;Q-B_4_1).

Count Row % Col % Total %	SOLE OWNER	HAVE PARTNER/S	Row Total %
GENERAL	22 75.9 30.1 22.9	7 24.1 30.4 7.3	29 30.2
SPECIAL	10 71.4 13.7 10.4	4 28.6 17.4 4.2	14 14.6
LAB SUB	16 69.6 21.9 16.7	7 30.4 30.4 7.3	23 24.0
L&M SUB	25 83.3 34.2 26.0	5 16.7 21.7 5.2	30 31.3
Column Total %	73 76.0	23 24.0	96 100.0

Refer to question B_4 of the questionnaire for the original question

Table 5.5 - Ownership of Firms

5.3 PERSONAL BACKGROUND OF OWNER

This section describes the background of the firm owner. Aspects covered include: age; schooling; tertiary education; and literacy. Details of type of training and experience gained while employed prior to starting the firm are included in the section 5.4.

5.3.1 Age of Owner

In the following table (Table 5.6), the ages of firm owners are categorised in ten-year increments.

Count Row %	20 TO 29 YRS OLD	30 TO 39 YRS OLD	40 TO 49 YRS OLD	50 TO 59 YRS OLD	60 TO 69 YRS OLD	Row Total %
ALL	16 16.7	35 36.5	26 27.1	13 13.5	6 6.3	96 100.0

Refer to question C_1 of the questionnaire for the original question

Table 5.6 - Age of Firm Owner

Individuals' ages ranged from 20 to 67 years, while the average age for the sample was between 40 and 41

(pg.25/AppH;Q-C_1). The majority of contractors (16.7% + 36.5% = 53.2%) were younger than 40, b
most of these were between 30 and 40 years old.

5.3.2   Father's Occupation

Contrary to what was expected, the majority of the respondents reported that their fathers were/had not
been employed in the building industry or were/had not been businessmen. This can be seen from the da
(pg.26/AppH;Q-C_2). Responses to this question came from 93 interviewees, where 32 (34.4%) had fat
who were/had been employed in the building industry, 13 (14.0%) had fathers who were/had been
businessmen, and 48 (51.6%) had fathers who were/had been employed in an occupation other than thes
. two.

Two reasons might explain the above findings. Firstly, the effect of the restrictive legislation (discussed
section 2.5.2) which prevented Blacks from rising above general worker level in the building industry is
likely to have impacted on the offspring of these workers - by deterring them from following in their
fathers' footsteps. Secondly, the traditional tribal manner of procuring shelter (that is, the family or tribe
rather than a building contractor, built the shelters) might have influenced the career aspirations of
individuals to whom this was normal. In other words, if the practice of independent building contracting
was not a part of the culture in which an individual was raised, then it is less likely that the individual
would consider building contracting as a career. This might apply to recently urbanised families and
individuals.

5.3.3   Educational Background of Firm Owners

The following table, Table 5.7, reflects the highest school standard passed by the interviewees.

There are seven junior school standards (sub A, sub B, and standards 1 to 5) and five senior school
standards (standards 6 to 10). Column headings indicate the highest standard passed and assume that all
lower standards were passed.

The majority of the interviewees (96.8%) had completed junior school, but only 34.7% had completed
senior school. A large number (51.6%) had left senior school prematurely - with no higher than a standa
8 education. A higher percentage (50.0%) of specialist sub-contractors had passed matric than was the ca
for any of the other types of firm owner. This result was expected, given the more academic nature, and
therefore higher entry requirements, of the study of electricity compared with other building trade
syllabuses. On the whole, these data support Krafchik's (1990:90-91) finding that about 95% of sub-
contractors are *functionally literate* (minimum of five years schooling). Notably, however, 100% of gene

contractors were functionally literate (see Table 5.5). See also section 5.3.6 for more detail on literacy.

Count Row % Col % Total %	SUB B	STD 5	STD 6	STD 8	STD 9	STD 10 (MATRIC)	Row Total %
GENERAL	0 0.0 0.0 0.0	3 10.7 33.3 3.2	5 17.9 41.7 5.3	7 25.0 28.0 7.4	2 7.1 15.4 2.1	11 39.3 33.3 11.6	28 29.5
SPECIAL	1 7.1 33.3 1.1	0 0.0 0.0 0.0	2 14.3 16.7 2.1	6 21.4 12.0 3.2	1 7.1 7.7 1.1	7 50.0 21.2 7.4	14 14.6
LAB SUB	1 4.3 33.3 1.1	4 17.4 44.4 4.2	2 8.7 16.7 2.1	3 26.1 24.0 6.3	6 26.1 46.2 6.3	4 17.4 12.1 4.2	23 24.0
L&M SUB	1 3.3 33.3 1.1	2 6.7 22.2 2.1	3 10.0 25.0 3.2	9 30.0 36.0 9.5	4 13.3 30.8 4.2	11 36.7 33.3 11.6	30 31.3
Column Total %	3 3.3	9 9.5	12 12.6	25 26.3	13 13.7	33 34.7	95 100.0

Refer to question C_3 of the questionnaire for the original question

Table 5.7 - Educational Background of Firm Owners (Highest Standard Passed)

5.3.4 Technical Schooling and Tertiary Education

Although it is assumed that most readers will be conversant with the South African technical and tertiary education system, Appendix A has been included to explain some necessary detail. For example, the data in Table 5.8 will probably make more sense if studied in conjunction with this appendix.

Seventy-four (74) respondents answered the question (question C_4) on whether or not they had obtained a qualification from a college, Technikon or university. Of these, 26 (35.1%) indicated that they had obtained such a qualification (pg.28/AppH;Q-C_4). Considering, however, that these 26 firms formed only 27% of the total survey sample, it is clear that the majority of firm owners had not obtained formal trade-oriented education. Thus, it is argued that the majority of respondents had not taken steps to establish a foundation of theoretical knowledge for their careers as building contractors. Indeed, as will be seen in sections 5.4.3 and 5.4.4 of this analysis, this lack of formal training later causes management problems for individuals, evidenced by the high demand for training among owners of established firms.

These data tend to further support that building contracting is an easy vocation to enter and that on-the-job training is an important part of SSCE support.

94

Count	NO. (ALL TYPES OF CONTRACTORS)	COLUMN %
National Certificate in Bricklaying	2	7.7
National Certificate in Building Science	2	7.7
N2 National Certificate in Carpentry	1	3.8
N3 National Certificate	2	7.7
N3 in Bricklaying	3	11.5
N4 at Technikon	3	11.5
N6 Civil Engineering and Building	2	7.7
National Technical Certificate (NTC 3) in Electrical Engineering	1	3.8
National Technical Certificate (NTC 6) in Business Theory	1	3.8
Diploma in Bricklaying	1	3.8
Diploma in Business Administration	1	3.8
Diploma in Taxation (University of Witwatersrand)	1	3.8
Electrical Engineering	1	3.8
Degree in Civil Engineering	1	3.8
L.L.B. (Law) Business Economics	1	3.8
No response	3	11.5
Column Total / %	26	100.0

Refer to question C_4_1 of the questionnaire for the original question

Table 5.8 – Type of Qualification Obtained From a College, Technikon or University

Table 5.8 contains the responses of those individuals who had obtained qualifications - noting the type of qualification obtained. Although these qualifications tended to be predominately technical in nature, there were, surprisingly, a few (3?) respondents who held university degrees.

5.3.5 Other Sources of Income

As was expected, the majority of firm owners (87.5%) cited building contracting as their main source of income (pg.30/AppH;Q-C_5). This indicates that relatively few respondents are likely to have entered self employed contracting via the 'commercial route' defined in section 2.4.1 (b).

5.3.6 Literacy

Levels of literacy among the respondents could not be established from the survey data. However, it was found that only 13.5% of the respondents cited English as the language that they speak, read and write be (pgs.31-32/AppH;Q-C_6_1 & C_6_2). Notably, none of the respondents indicated that Afrikaans was the best language. English was, however, the most commonly reported as the second best language.

These findings were expected and are important. Considering the finding (in Table 5.7) that the majority the interviewees did not complete their secondary schooling, it is likely that these contractors would experience great difficulty operating in a predominately English- or Afrikaans-speaking business environment. Indeed, the model documents (contracts, *etc.*) produced by BIFSA are only available in English or Afrikaans. The likely effect of this constraint on black contractors would be that, although they might readily find black Zulu-speaking clients in their own communities, they would be considerably less able to establish themselves in the formal building industry. This notion is supported by the findings (reported in section 5.7.2) that the firms' operated mainly in their own communities (black townships).

5.4.   TRAINING

This section deals with the training background of the interviewees. Training is taken to include on-the-job training and details are included of the type and duration of experience obtained whilst employed by a building contractor. In addition, aspects such as: apprenticeship; adequacy of training; demand for further training and type of training required; and awareness of and preference for training institutions; are included.

5.4.1   History of Employment in the Building Industry

Although the majority (76%) of the firm owners had been employees in the building industry prior to starting their own firms, a significant 24% had not (pg.33/AppH;Q-D_1).

In Table 5.9, the column marked 'N/A' reflects numbers of the 96 respondents who had never been employed in the capacity described in the column to the left.

Twenty-three (23) (24.0%) respondents had been unskilled employees for various durations. As can be seen from the table, the majority had been unskilled employees for less than 4 years. Notably, most of these had been employed for less than one year.

Twenty-five (25) (26.0%) respondents had been semi-skilled employees. The majority of this group had been semi-skilled employees for less than 4 years. Just over half of this sub-group had been employed for more than 1, but less than 4 years, while just under half had been employed for less than 1 year.

Thirty-four (34) (35.4%) respondents had been skilled employees. The majority of this group had obtained less than 4 years of skilled experience. A large majority of this sub-group had been employed for more than 1, but less than 4 years. Only 7 (7.3%) respondents had been employed as foremen. Three (3) of these 7 had been employed as foremen for less than 2 years, while only 1 had been employed as a foreman for more than 10 years. Four (4) (4.2%) respondents had been employed as clerks. None of these had been employed in a clerical position for longer than 6 years.

Count Row %	N/A	LESS THAN 1 YR	MORE THAN 1 YR BUT LESS THAN 2 YRS	MORE THAN 2 YRS BUT LESS THAN 4 YRS	MORE THAN 4 YRS BUT LESS THAN 6 YRS	MORE THAN 6 YRS BUT LESS THAN 8 YRS	MORE THAN 8 YRS BUT LESS THAN 10 YRS	MORE THAN 10 YRS	Row Total %
UNSKILLED	73 76.0	8 8.3	1 1.0	6 6.3	2 2.1	2 2.1	2 2.1	2 2.1	96 100.0
SEMI-SKILLED	71 74.0	8 8.3	4 4.2	5 5.2	1 1.0	1 1.0	2 2.1	4 4.2	96 100.0
SKILLED	62 74.0	3 3.1	9 9.4	8 8.3	4 4.2	1 1.0	1 1.0	8 8.3	96 100.0
FOREMAN	89 92.7	2 2.1	1 1.0	0 0.0	2 2.1	1 1.0	0 0.0	1 1.0	96 100.0
CLERICAL	92 95.8	1 1.0	0 0.0	2 2.1	1 1.0	0 0.0	0 0.0	0 0.0	96 100.0
MANAGEMENT	91 94.8	0 0.0	1 1.0	3 3.1	1 1.0	0 0.0	0 0.0	0 0.0	95 100.0

Refer to question D_1_1 of the questionnaire for the original question.

These data apply to ALL types of contractors

Table 5.9 - Duration of Employment in the Building Industry

Only 5 (5.2%) respondents had been employed in management positions. Four (4) of these had been employed for less than 4 years, while 1 had been employed for between 4 and 6 years.

Table 5.9 thus illustrates two main points. Firstly, there was a distinct lack of management-related experience (foreman, clerical or management) in the employment history of the sample. Secondly, the duration of employment in any capacity, tended to be short - mostly under 4 years.

In Table 5.10 the extent to which respondents had progressed through the employment ranks can be see Data for clerical and management employees are excluded, since these positions would not necessarily b filled by individuals who had previously been tradesman or workers. The most striking fact to emerge fi this data is that the majority (14.6 + 16.7 + 26.0 + 3.1 = 60.4%) had only ever been employed in a single capacity. Thus, only 13.5% had been employed in more than one capacity. The combinations of these employment capacities are reflected in the column headings of the table. It stands to reason that th ideal employment background for an individual wishing to start a firm would have included unskilled, s skilled, skilled and supervisory (foreman) experience. Notably, only 1 respondent reported this to be the case.

Count / Row % / Col % / Total %	NIL	US	SS	SK	FMN	US AND SS	US AND SK	US AND FMN	SS AND SK	SS AND FMN	SK AND FMN	US AND SS AND SK	US AND SS AND FMN	US AND SK AND FMN	SS AND SK AND FMN	US AND SS AND SK AND FMN	Row Total %
GENERAL Count	10	5	3	6	1	0	1	1	1	1	0	1	0	0	0	0	29
Row %	34.5	17.2	10.3	20.7	3.4	0.0	3.4	3.4	3.4	3.4	0.0	3.4	0.0	0.0	0.0	0.0	30.2
Col %	40.0	35.7	18.8	24.0	33.3	0.0	100.0	50.0	33.3	33.3	0.0	33.3	0.0	0.0	0.0	0.0	
Total %	10.4	5.2	3.1	6.3	1.0	0.0	1.0	1.0	1.0	1.0	0.0	1.0	0.0	0.0	0.0	0.0	
SPECIAL Count	6	0	3	4	1	0	0	0	0	0	0	0	0	0	0	0	14
Row %	42.9	0.0	21.4	28.6	7.1	0.0	0.0	0.0	0.0	0.0	0.0	0.0	0.0	0.0	0.0	0.0	14.6
Col %	24.0	0.0	18.8	16.0	33.3	0.0	0.0	0.0	0.0	0.0	0.0	0.0	0.0	0.0	0.0	0.0	
Total %	6.3	0.0	3.1	4.2	1.0	0.0	0.0	0.0	0.0	0.0	0.0	0.0	0.0	0.0	0.0	0.0	
LAB SUB Count	3	6	6	5	0	1	0	1	0	0	0	1	0	0	0	0	23
Row %	13.0	26.1	26.1	21.7	0.0	4.3	0.0	4.3	0.0	0.0	0.0	4.3	0.0	0.0	0.0	0.0	24.0
Col %	12.0	42.9	37.5	20.0	0.0	50.0	0.0	50.0	0.0	0.0	0.0	33.3	0.0	0.0	0.0	0.0	
Total %	3.1	6.3	6.3	5.2	0.0	1.0	0.0	1.0	0.0	0.0	0.0	1.0	0.0	0.0	0.0	0.0	
LAM SUB Count	6	3	4	10	1	1	0	0	2	0	1	1	0	0	0	1	30
Row %	20.0	10.0	13.3	33.3	3.3	3.3	0.0	0.0	6.7	0.0	3.3	3.3	0.0	0.0	0.0	3.3	31.3
Col %	24.0	21.4	25.0	40.0	33.3	50.0	0.0	0.0	66.7	0.0	100.0	33.3	0.0	0.0	0.0	100.0	
Total %	6.3	3.1	4.2	10.4	1.0	1.0	0.0	0.0	2.1	0.0	1.0	1.0	0.0	0.0	0.0	1.0	
Column Count	25	14	16	25	3	2	1	2	3	0	1	3	0	0	0	1	96
Total %	26.0	14.6	16.7	26.0	3.1	2.1	1.0	2.1	3.1	0.0	1.0	3.1	0.0	0.0	0.0	1.0	100.0

Refer to question D_1_1 of the questionnaire for the original question

NEL	=	No building industry experience as an employer
US	=	Experience as an UNSKILLED EMPLOYEE in the building industry
SS	=	Experience as a SEMI-SKILLED EMPLOYEE in the building industry
SK	=	Experience as a SKILLED EMPLOYEE in the building industry
FMN	=	Experience as a FOREMAN EMPLOYEE in the building industry

N.B. Other column headings are combinations of the above

Table 5.10 - Type of Experience as an Employee

General contractors accounted for 40% of those who had never been employed in the building industry, while labour-only sub-contractors accounted for 12% of this group. Labour-only sub-contractors account for the largest percentage of the group who had only been unskilled employees (42.9%) as well as of the group who had only ever been semi-skilled employees (37.5%). General contractors made up the highest percentage of the group who had only ever been skilled employees (24.0%).

5.4.2 Apprenticeships

The following table (Table 5.11) reports the response concerning whether or not formal apprenticeships were served.

Count Row %	NEVER SERVED A FORMAL APPRENTICESHIP	SERVED A FORMAL APPRENTICESHIP IN ONE TRADE	SERVED A FORMAL APPRENTICESHIP IN TWO TRADES	Row Total %
ALL TYPES OF CONTRACTORS	68 70.8	28 20.8	8 8.3	96 100.0

Refer to question D_2_1 of the questionnaire for the original question

Table 5.11 - Extent of Apprentice Training

These data indicate clearly that a large majority (70.8%) of respondents had never served formal apprenticeships. It must be noted that in the original responses, five interviewees indicated that they had served two formal apprenticeships, namely Bricklaying and Plastering. There is a formally designated tra 'Bricklayer and Plasterer', and it was assumed that this is what was referred to. These five responses hav therefore been included under the column for those having served one apprenticeship.

5.4.3 Management Training

In response to question D_3, 42.7% of the respondents indicated that they had, prior to starting their ow firms, received formal education or training in how to manage a building firm (pg.49/AppH).

This result appears to be inaccurate, or exaggerated, in that it contradicts two other findings. In Table 5. only 3 respondents reported that they had obtained a business-related qualification from an educational institution. In addition, it was seen in Table 5.9 that only 8 respondents had obtained experience as employees in a management capacity.

5.4.4 Demand for Further Training

Question D_4 required interviewees to indicate whether or not they required further training. It should be noted that this question does not relate exclusively to the previous question, in that it provides for the response that skills training is required.

A large majority of respondents (87.5%) indicated that they required further training. This suggests that, even if the results of question D_3 were accurate, any previously acquired training had been perceived as inadequate.

The data in Table 5.12 highlight the fact that there is a considerable demand for all types of training. The greatest demand is for business management training, followed by production management training (managing the building process) and then trade skills training. That there is a demand for trade skills training is cause for concern, in that it indicates that firm owners undertake building work which they themselves do not feel sufficiently skilled to perform.

Count Row %	TRADE SKILLS	PRODUCTION MANAGEMENT	BUSINESS MANAGEMENT	PLAN DRAWING	QUANTITY SURVEYING	Row Total %
ALL	33 34.4	43 44.8	57 59.4	1 1.0	1 1.0	135 140.6

Refer to question D_4_1 of the questionnaire for the original question

N.B.   The question allowed for multiple responses. For this reason the row percentages in the above table (which are percentages of the total survey size of 96) exceed 100%.

Table 5.12 - Type of Further Training Required

A deficiency in trade skills must directly affect the quality of the built product, while a deficiency in management skills mainly affects the business, but may also affect the built product. The deficiency in trade skills, particularly where the firm owner is part of the team of 'skilled' workers, must be a major contributor to the generally poor reputation of black SSCEs. These data thus support the argument in section 2.5.3 that black SSCEs are disadvantaged by their generally poor reputation.

5.4.5 Steps Taken to Obtain Further Training

Only 18 of the 84 respondents (21.4%) who indicated that they required further training, had done anything about getting this training (pg.56/AppH;Q-D_4_2). This is a very important result, since it suggests, as argued in Chapter 4, that existing training institutions or courses are inaccessible to the majority of those who desire training.

Count / Col %	NUMBER	Column %
Applied to enrol	1	5.6
Applied for overseas scholarship	2	11.1
Applied to bank for loan - not yet granted	1	5.6
Applied for scholarship	1	5.6
Saving/saved money and intend to enrol	3	16.7
Enroled on a business management course	2	11.1
Applied to African Builders' Association for a bursary	1	5.6
Completed a one-day course in business management at Kwazulu Training Trust	1	5.6
Attended a short crash course in business management	1	5.6
Enroled on a course, but dropped out due to the demands of the business	2	11.1
Made enquiries about enroling	1	5.6
Attended night school	1	5.6
No response	1	5.6
Column Total %	18	100.0

Refer to question D_4_2 of the questionnaire for the original question

Table 5.13 - Action Taken Towards Obtaining Further Training

Although respondents were not asked why they had not done anything about obtaining the training they desire, it is likely that this is due to two main reasons - lack of time, and expense. This speculation is based on the fact that, as can be noted in Table 5.13, the majority of those who did do something about getting further training either attended short courses, applied for financial assistance, or began saving towards the cost of the training. Only 5 of the 18 respondents (27.8%) indicated that they had actually enroled for and completed courses.

5.4.6 Preferred Training Institutions

Question D_4_4 was included to determine whether or not there was a demand for (or an awareness of) the many training courses specifically aimed at SSCEs (that is, those described in Appendix F and evaluated in Chapter 4). Only three of these institutions were mentioned by name - KTT, SBDC and BIFSA. The actual responses to the question are presented below in Table 5.14. These data confirm the argument made in Chapter 4 that the existing training infrastructure is ineffective. Further, the data endorse the hypothesis made in Chapter 4 that BIFSA training is inappropriate for the average SSCE. Indeed, the data also tend to confirm the argument that the risk-minimising tendencies of these institutions (the "A-Grade" contactor syndrome) effectively filter out virtually all SSCEs.

The majority (80.0%) of the responses contained in Table 5.14 can be abstracted into the following three categories:-

(a) Technikons and Technical Colleges: These institutions offer nationally recognised certificates and diplomas. Technikons and Technical Colleges, respectively, accounted for 29.4% and 4.7% of the responses (from 85 respondents). This type of institution is probably favoured because a widely recognised qualification can be obtained, and because part-time courses are presented.

Count Col %	NUMBER	Column %
KTT	5	5.8
Technikon	24	28.2
Technical College	4	4.7
Pinetown Training Centre (included with KTT)	0	0
Technikon (because it presents part-time courses)	1	1.2
BIFSA	1	1.2
Damelin Correspondence College	8	9.4
University	1	1.2
Anywhere	3	3.5
Murray & Roberts (a large building contracting firm)	2	2.3
Electrical Contractors Association	1	1.2
Any correspondence school	1	1.2
Executive College	2	2.3
Lyceum College	1	1.2
George Tabor College	2	2.3
Chamdor Training Institute	3	3.5
Department of Manpower	2	2.3
School of Business Leadership	1	1.2
SBDC	1	1.2
Birnham College	2	2.3
Training College	2	2.3
No response (don't know)	19	22.4
Column Total %	86	101.1

Refer to question D_4_4 of the questionnaire for the original question.

N.B.   The question allowed for multiple responses. For this reason the column percentages in the above table (which are percentages of the total sample size of 85) exceed 100%.

Table 5.14 - Preferred Training Institutions

(b) Correspondence schools: Seventeen (17) respondents (20.0% of the 85) named correspondence colleges as where they would go for training. Presumably this is because they offer courses which do not require the firm owner to leave his business unattended during working hours.

(c) Unspecified: Twenty-two (22) of the interviewees (25.9% of the 85) either did not know where they would go for further training, or did not name a specific institution. This suggests that training institutions do not do enough to publicise their courses among the community of SSCEs.

The following table (Table 5.15), summarises what the firm or firm owner was doing during every quarter of 1990 and 1991. The majority of firms were building for their own clients during these two years. Notably, there was a gradual increase in the number of firms that were active after the second quarter of 1990. This may indicate that the usual markets in which the firms operated were improving, or that the firms were generally improving at successfully securing contracts.

Count Row %	FIRM HAD WORK	FIRM HELPED ANOTHER FIRM	REVERTED TO EMPLOYEE STATUS	NOT ACTIVE - SOUGHT WORK	LEFT BUILDING INDUSTRY	OTHER	Row Total %
JAN TO MARCH 1990	73 76.0	7 7.3	1 1.0	8 8.3	0 0.0	7 7.3	96 100.0
APRIL TO JUNE 1990	64 66.7	12 12.5	3 3.1	9 9.4	1 1.0	7 7.3	96 100.0
JULY TO SEPT 1990	69 71.9	5 5.2	2 2.1	12 12.5	0 0.0	8 8.3	96 100.0
OCT TO DEC 1990	70 72.9	13 13.5	3 3.1	7 7.3	1 1.0	2 2.1	96 100.0
JAN TO MARCH 1991	71 74.0	10 10.4	0 0.0	11 11.5	0 0.0	4 4.2	96 100.0
APRIL TO JUNE 1991	71 74.0	12 12.5	0 0.0	6 6.3	1 1.0	6 6.3	96 100.0
JULY TO SEPT 1991	78 81.3	6 6.3	1 1.0	7 7.3	0 0.0	4 4.2	96 100.0
OCT TO DEC 1991	81 84.4	5 5.2	0 0.0	6 6.3	1 1.0	3 3.1	96 100.0

Refer to Module E of the questionnaire for the original question

Table 5.15 - Activities of Firms During 1990 and 1991

The data in columns three to seven of the table detail what firms or their owners were doing when they were not building for their own clients. The two most common activities were, respectively, the firm assisted another firm and the firm sought work. The third most common activity of firms that had no work was 'other'. This refers to something 'other' than assisting another firm, reverting to employee status, seeking work, or leaving the building industry. The only obvious exception to this range of options is that the firm was not active, but did not seek work. It cannot, however, be assumed that this is what was meant by respondents who answered 'other'. The fourth most common activity reported, was that firm owners reverted to employee status. It must be noted though, that this was a considerably less common response than those mentioned above. The least common activity reported, was that firm owners left the building industry.

The data representing firms who had their own work during 1990 and 1991 could be misleading, in that the totals do not represent exactly the same firms. Table 5.16 is included to clarify this problem.

Count Row % Col % Total %	FIRM NOT ACTIVE FOR WHOLE OF 1990 AND 1991	FIRM ACTIVE FOR WHOLE OF 1990 AND 1991	Row Total %
GENERAL	21 72.4 33.3 21.9	8 27.6 24.2 8.3	29 30.2
SPECIAL	9 64.3 14.3 9.4	5 35.7 15.2 5.2	14 14.6
LAB SUB	14 60.9 22.2 14.6	9 39.1 27.3 9.4	23 24.0
L&M SUB	19 63.3 30.2 19.8	11 36.7 33.3 11.5	30 31.3
Column Total %	63 65.6	33 34.4	96 100.0

Refer to Module E of the questionnaire for the original question

Table 5.16 - Continuous Activity During 1990 and 1991

From Table 5.16 it can be seen that only 34.4% of the firms were active during every quarter of 1990 and 1991. These data confirm the suggestion made in section 2.4.3 that, as demand drops, which was the case in the period under consideration, SSCEs experience long periods of inactivity. Of this 34.4%, 33.3% were labour-and-material sub-contractors, 27.3% were labour-only sub-contractors, 24.2% were general contractors and 15.2% were specialist sub-contractors.

Considered as separate groups, 39.1% of all labour-only sub-contractors, 36.7% of all labour-and-material sub-contractors, 35.7% of all specialist sub-contractors and 27.6% of all general contractors, were active in every quarter of both years. Notably, general contractors were more prone to periods of inactivity than the other groups.

5.6   ENTRY INTO BUILDING INDUSTRY

In this section, factors surrounding the entry into the building industry of the firms are described. Aspect covered include: motivation for starting the firm; role of the owner's previous employer; how work was first obtained; and most serious problems encountered in establishing the firm.

5.6.1   Reason for Entering the Industry as a Contractor

In the following table, (Table 5.17), data are presented which describes what motivated individuals to establish their own contracting firms.

Count Row %	FAMILY	COULD NOT FIND WORK	INDEPEN-DENCE	THOUGHT INCOME WOULD BE BETTER	LACK OF EDUCA-TION	AMBITION	EXPLOI-TATION	PREVIOUS EMPLOYER WENT INSOLVENT	Row Total %
ALL TYPES OF CONTRACTOR	9 9.5	20 21.1	23 24.2	16 16.8	2 2.1	23 24.2	1 1.1	1 1.1	95 100.0

Refer to question F_1 of the questionnaire for the original question

Table 5.17 - Reasons for Becoming a Contractor

The results in Table 5.17 can be divided into two categories - positive and negative reasons. Positive reasons include: entering or inheriting a family business; the desire for independence; and ambition. Negative reasons include: unemployment; perceived increased earnings; lack of education for other kind of work; exploitation by previous employer; and insolvency of previous employer (retrenchment). Note that Krafchik (1990:87) describes 'perceived increased earnings' as a positive reason for entry. But, since the majority of respondents actually left their employment with insufficient experience (see Table 5.10), it is argued that they were (prematurely) 'pushed', rather than 'pulled' into self-employment. This has therefore been classified as a negative reason for entry.

Although there were fewer positive motivators than negative, the majority (57.9%) of the respondents have started their firms for positive reasons. The most common positive motivators were ambition (24.2%) and the desire for independence (24.2%).

Unemployment (21.1%) and the perception that earnings would improve (16.8%) were the most common negative motivators.

5.6.2   Role of Previous Employer

Previous employers were not a major source of assistance to the new firm owners. Only 18.8% of these indicated that they had received assistance from their previous employer at start-up (pg.109/AppH;Q-F_2). But, in every case the previous employer was involved in the building industry (pg.110/AppH;Q-F_2_1).

In Table 5.18 below, details are given of the nature of the assistance that previous employers provided.

Count Row %	EXTENDED A LOAN	PASSED ON EXCESS WORK	PASSED ON CONTACTS WHICH RESULTED IN WORK	LENT EQUIPMENT OR TOOLS	Row Total %
ALL	4 22.2	11 61.1	14 77.8	2 11.1	31 172.2

Refer to question F_2_2 of the questionnaire for the original question

N.B.   The question allowed for multiple responses. For this reason the row percentages in the above table (which are percentages of the sample size of 18 respondents who had received assistance from their previous employers), exceed 100%.

Table 5.18 - Nature of Assistance Initially Provided by Previous Employer

The most common types of assistance provided by previous employers were: passing on contacts; and passing on excess work. Notably, previous employers were an insignificant source of finance and building equipment.

As was expected, previous employers were found to reduce or terminate support once the firm had been established. This is seen in the fact that only 8.3% of the respondents reported that they were still receiving support from this source (pg.115/AppH;Q-F_3). Table 5.19 below, shows the nature of this sustained support.

Count Row %	EXTENDED A LOAN	PASSED ON EXCESS WORK	PASSED ON CONTACTS WHICH RESULTED IN WORK	LENT EQUIPMENT OR TOOLS	Row Total %
ALL	2 25.0	5 62.5	6 75.0	0 0.0	13 162.5

Refer to question F_3_1 of the questionnaire for the original question

N.B.   The question allowed for multiple responses. For this reason the row percentages in the above table (which are percentages of the sample size of 8 respondents who had received ongoing assistance from their previous employers), exceed 100%.

Table 5.19 - Nature of Assistance Currently (1992) Provided by Previous Employer

The data in Table 5.19 mirror the findings in Table 5.18, but for a smaller number of respondents. Without intending to trivialise the attempts made by previous employers to assist the new and/or established firms, it must be noted that the support they provided tended to cost them little by way of time or effort. Indeed, as

will be seen from Table 5.21, the assistance fell short of improving the most serious problems experience
by the firms.

5.6.3   Source of First Contract

Table 5.20 below gives the sources of the firms' first contracts, where these were not acquired through
previous employers.

Client Col %	NUMBER
Client responded to advertisement - firm owner's house was used as proof of ability	1 1.9
Client approached the firm owner after seeing the quality of work on his own house	12 23.1
Client responded to advertisement - by word of mouth or with pamphlets or in the press	13 25.0
Employed by a friend	11 21.1
Employed by a relative	2 3.8
Employed by a contractor or developer	11 21.1
Client was found by a current or previous partner	2 3.8
Client was found through Kwazulu Training Trust	1 1.9
Client was found through Kwazulu Finance and Investment Corporation	2 3.8
Employed by client to finish off work abandoned by another contractor	1 1.9
Client was found through Small Business Development Corporation	1 1.9
Column Total %	57 109.0

Refer to question F_4 of the questionnaire for the original question

N.B.   Five (5) respondents named 2 sources from which their first contracts were obtained. For this reason the
column percentages in the above table (which are percentages of the total sample size of 52), exceed 100%.

Table 5.20 - Source of First Contract (other than previous employer)

In Table 5.20 it can be seen that advertisement (25%), in one form or another, was successfully used as
method of attracting the firms' first clients. It is interesting to note that the second most common source
work was clients who approached the firm owner after seeing the quality of his work on his own house
(23.1%). Friends and building contractors also emerged as important sources of first contract, although i
not clear which party made the approach.

Count Col %	MOST SERIOUS PROBLEM IN STARTING FIRM	Col %	SECOND MOST SERIOUS PROBLEM IN STARTING FIRM	Col %
Clients had no confidence in firm's ability	1	1.0	N/A	
Lack of finance/working capital	58	60.4	14	14.6
Could not find clients	6	6.3	3	3.1
Lack of builders equipment	6	6.3	11	11.5
Material	N/A		13	13.5
Could not advertise	N/A		2	2.1
Transport	3	3.1	10	10.4
Could not make a profit	1	1.0	1	1.0
Labour/labourers/employees	2	2.1	7	7.3
Slow payment by clients/bad debts	1	1.0	2	1.1
Inadequate security against theft	1	1.0	2	2.1
Unable to find labourers	2	2.1	1	1.0
Clients	1	1.0	3	3.1
Lack of a workshop	N/A		1	1.0
Lack of technical skill	1	1.0	2	2.1
People prefer white contractors	1	1.0	N/A	
Unable to find sub-contractors	1	1.0	N/A	
Industrial Council inspectors	1	1.0	N/A	
Equipment suppliers	N/A		1	1.0
Communication	N/A		1	1.0
Fear of complaints about quality of work	N/A		1	1.0
Cost of labour	N/A		1	1.0
Having to accept "take-it-or-leave-it" offers	1	1.0	N/A	
Township riots	N/A		1	1.0
Lack of business/financial management expertise	1	1.0	2	2.1
Administering jobs	N/A		1	1.0
Dishonesty of clients	N/A		1	1.0
Unreliability of labourers	1	1.0	N/A	
Repayment of loans	1	1.0	1	1.0
Shortage of skilled labour	N/A		1	1.0
Competition	N/A		1	1.0
Relationship with main contractor	1	1.0	N/A	
Cost of materials	N/A		2	2.1
Relationship with sub-contractors	1	1.0	N/A	
No response	4	4.2	10	10.4
Column	Total %	96	100.0	96	100.0

Refer to question F_5 of the questionnaire for the original question

Table 5.21 - Two Most Serious Problems Encountered in Starting Firm

The actual responses, regarding the most serious problems encountered when the firm was established, are presented above in Table 5.21. This was done because it is not clear what some of the responses were intended to mean. Nevertheless, it is clear that 'access to finance or working capital' was by far the most serious problem (reported by 60.4% of the respondents). 'Finding clients' and 'access to equipment' were

the next two most commonly cited serious problems, but, in each case, these were reported by a relative
small 6.3% of the respondents.

Significantly, 'access to finance or working capital' was also the most commonly reported (by 14.6% of
respondents) second most serious problem. Other commonly cited second most serious problems include
'material' (13.5%) - although it is not clear whether this refers to the cost of materials or to materials
supply problems; 'lack of builders equipment' (11.5%); and 'transport' (10.4%).

These data are largely consistent with the external constraints described in section 2.5.2

5.7 ACHIEVEMENTS AND ACTIVITIES OF THE FIRMS

This section describes the firms' activities. Aspects covered include: type of client; how contract amoun
are established; number of contracts undertaken per year; comparison of first and last contracts; monthly
turnover; profit; where firms operate; current utilisation of firms' resources; and how success is measur

5.7.1 Type of Client

The following two tables, (Tables 5.22 and 5.23) contain the responses to questions G_1 and G_2,
respectively.

Count Row %	ALWAYS	SOMETIMES	NEVER	Row Tot
Employment by **INDIVIDUALS / HOME OWNERS** since January 1990	21 72.4	8 27.6	0 0.0	
Employment by **ORGANISATIONS / COMPANIES** since January 1990	1 3.4	13 44.8	15 51.7	
Employment by **CHURCHES** since January 1990	0 0.0	1 3.4	28 96.6	

Refer to question G_1 of the original questionnaire for the original question

Table 5.22 - Type of Client : General Contractors

From Table 5.22 it can clearly be seen that, during 1990 and 1991, general contractors were most
frequently employed by individuals, rather than companies. These data apply to both types of general
contractor (that is, mainly new work and mainly alterations and additions work).

Client Row %	NO RESPONSE	ALWAYS	SOMETIMES	NEVER	Row Total %
Employment by DEVELOPERS since January 1990	6 9.0	7 10.4	29 43.3	25 37.3	67 100.0
Employment by LARGE CONTRACTORS since January 1990	6 9.0	7 10.4	24 35.8	30 44.8	67 100.0
Employment by SMALL AND MEDIUM CONTRACTORS since Jan 1990	6 9.0	9 13.4	39 58.2	13 19.4	67 100.0
Employment by INDIVIDUALS / OWNERS since January 1990	5 7.5	37 55.2	22 32.8	3 4.4	67 100.0

Refer to question G_2 of the questionnaire for the original question

Table 5.23 - Type of Client : Sub-contractors

The data in Table 5.23 show that, as was the case for general contractors, sub-contractors are most frequently employed by individuals. However, a considerable number of sub-contractors reported that they 'sometimes' work for developers or contractors. This latter finding was expected, since the most likely employer of a sub-contractor would be a contractor.

5.7.2 Areas in which Work was Obtained

The responses to question G_7 indicate the areas in which clients commissioned work (pgs.173-176/AppH). The majority of firms always (50.0%) or sometimes (36.5%) obtained work in their own communities. Similarly 39.6% of the firms noted that they were always employed by clients living in the townships, while 54.2% sometimes did so.

Clients living in cities were a minor source of work. This is seen in the responses that 46.9% of the firms had never worked for this type of client and 43.8% had only sometimes done so. Another minor source of work was clients living in the rural areas - 13.5% of the firms indicated that they sometimes, or always obtained work from this source (see Appendix F.B.3 for an explanation of why rural areas might form part of the metropolitan market)

On balance, though, these data taken together with those in Tables 5.22 and 5.23 support the suggestion made in section 2.5.2 (a) that black SSCEs are largely locked into the black residential market, contracting directly with owners.

5.7.3   Methods of Tendering for Work

The contents of Table 5.24 below, present the results of question G_3, in which respondents were asked
indicate how and how frequently they tender for work, negotiate contracts, or succumb to fixed offers fr
clients.

The results in Table 5.24 reveal important information regarding how, or indeed, whether or not, contra
values are established. The fairest method of attracting a quote from a contractor is to allow him to
determine his costs, add a mark-up and submit this total as a tender. In order to ensure that this tender is
not unduly inflated, it is common to invite other contractors to submit tenders. Notwithstanding Wells'
(1986:72-73) caveat that competitive tendering fails to produce the desired results, it is still the most
prevalent system in the South African formal building industry.

Count Row %	ALWAYS	SOMETIMES	NEVER	Row Total
How often is work acquired through a COMPETITIVE TENDER?	16 16.7	54 56.3	26 27.1	
How often is work acquired through NEGOTIATION?	57 59.4	33 34.4	6 6.3	
How often is work acquired through a TAKE-IT-OR-LEAVE-IT OFFER?	4 4.2	33 34.4	59 61.5	

Refer to question G_3 of the questionnaire for the original question

Table 5.24 - Methods of Tendering for/Acquiring Work

It is interesting to note that only 16.7% of the respondents in this survey always tender in competition wi
other contractors, 27.1% never do so, while 56.3% sometimes do so. Negotiation was reported to be the
most common method of agreeing upon a contract amount - 59.4% of the respondents always negotiate a
34.4% sometimes do so.

Although 61.5% of the firms reported that they never accept take-it-or-leave-it offers from their clients, a
alarmingly high 34.4% reported that they sometimes do so, while 4.2% always do so. Given that such
offers simply represent what the client is prepared to pay, rather than the correct price for the work, it is
likely that 38.6% of the respondents sometimes or always enter into contracts on which they are probably
going to make losses or, at best, break even. These data support the observations of Krafchik (1990),
Merrifield (1992) and Motlanthe (1990) that sub-contracting occurs under exploitive conditions (see secti
2.3.3, 2.4.2 and 4.1)

5.7.4   Number of Contracts Undertaken During 1991

In the table below (Table 5.25), the number of contracts undertaken by each firm during 1991 can be seen.
It seems unlikely that any firm would have undertaken and completed more than 60 contracts in a year as
indicated in the last column of the table, but this may be attributable to the fact that certain types of sub-
contracting firms generally undertake contracts of very short duration. In addition, it is possible that some
respondents may have counted each house (in a development of many houses) as a single contract, rather
than counting the whole development as one contract. The data should therefore be treated with some
caution.

Client Row %	ZERO OR DON'T KNOW OR NO RESPONSE	1 - 10	11 - 20	21 - 30	31 - 40	41 - 50	51 - 60	61 - 70	MORE THAN 70	Row Total %
ALL	6 6.3	50 52.1	24 25.0	7 7.3	2 2.1	3 3.1	0 0.0	1 1.0	3 3.1	96 100.0

Refer to question G_4 of the questionnaire for the original question

Table 5.25 - Number of Contracts Undertaken During 1991

The average number of contracts undertaken by the respondents during 1991 was 14.9 (computed from
pg.136/AppH). The most important observation that can be made from the data in Table 5.25 is that 52.1%
of the firms had undertaken fewer than 11 contracts and 77.1% had undertaken fewer than 21 contracts. A
comparison of these results and those in Table 5.43 indicates that contractors are unable to handle many
jobs simultaneously, rather than a lack of demand for their services. More importantly, however, the data
confirm Merrifield's (1992) observation that black SSCEs lack the capacity to compete with the larger white
developers (see section 2.5.2 f).

5.7.5   Comparison of First and Most Recent Contracts

(a) Type of employer: It was established from Tables 5.22 and 5.23 that, for all types of contractors, the
most common type of employer was the individual or owner. In Table 5.26, a comparison is made between
the type of client that employed the firm on both its first and most recent contracts. The purpose of this
comparison is to determine whether or not there had been a change in type of client, once the firm had
become established.

In addition to confirming the previous finding that the most common type of employer was the individual or
owner, the data in Table 5.26 reveal two important results. Firstly, there was little change in type of
employer between first and most recent contracts.

Count Column %		VERY FIRST JOB	MOST RECENT J...
CLIENT TYPE :	OWNER / INDIVIDUAL	77 80.2	6...
CLIENT TYPE :	DEVELOPER	11 11.5	1...
CLIENT TYPE :	SERVICE ORGANISATION (e.g. SAHT; RODECO; Urban Foundation HUC, etc.)	0 0.0	
CLIENT TYPE :	ZULU GOVERNMENT	0 0.0	
NO RESPONSE		8 8.3	
Column Total %		96 100.0	...

Refer to questions G_5 and G_6 of the questionnaire for the original questions

Table 5.26 - Comparison Between Type of Employer on First and Most Recent Contracts

Secondly, where there was a change, it involved an increase (from zero) in clients who were service
organisations or (homeland) government. The importance of the latter point is that it confirms the
contention made in Chapter 4 that the service organisations avoid employing contractors with no track
record. However, even once firms had become more established, this type of client was not a major
employer - only 10.4% of the respondents indicated having ever worked for such organisations.

(b) Type of building: In Table 5.27, a comparison is made between the type of building erected by the firm
on both its first and most recent contracts.

Count Column %	VERY FIRST JOB	MOST RECENT JOB
BUILDING TYPE : COMMERCIAL	13 13.5	15 15.6
BUILDING TYPE : RESIDENTIAL	71 74.0	68 70.8
BUILDING TYPE : SERVICE	4 4.2	5 5.2
NO RESPONSE	8 8.3	8 8.3
Column Total %	96 100.0	96 100.0

Refer to questions G_5_1 and G_6_1 of the questionnaire for the original questions

Table 5.27 - Comparison Between Type of Building Constructed on First and Most Recent Contracts

The vast majority of respondents indicated that they started out by building houses, and that they had
remained in this market. It can, however, be seen from Table 5.27 that there was a very slight tendency

(away from residential buildings) towards both commercial and service buildings. Nevertheless, these data confirm the points made in section 2.33 that SSCEs are well suited to (if not only capable of) the construction of small works and housing.

(c) Monthly turnover: Since many firms were periodically inactive, and that to some extent this is usual in the building industry, it was decided that annual turnover would be an inadequate indicator of growth. Rather, a comparison is made between monthly turnover on first and most recent contracts, but only for the number of months that the firm was working on any contract. The contact amount is divided by the contract period to give a monthly turnover for the duration of the contract. These results are displayed in the following two tables (Tables 5.28 and 5.29). The values in Table 5.28 are 1992 values, adjusted using the Consumer Price Index. The values in Table 5.29 are taken to be 1992 values, on the assumption that the most recent contract was undertaken in 1992.

Count Row % Col % Total %	LESS THAN R2 000	R2 001 TO R4 000	R4 001 TO R6 000	R6 001 TO R8 000	R8 001 TO R10 000	R10 001 TO R15 000	R15 001 TO R25 000	R25 001 TO R35 000	R35 001 TO R60 000	MORE THAN R60 000	Row Total %
GENERAL	1 3.8 9.1 1.1	2 7.7 12.5 2.3	1 3.8 11.1 1.1	2 7.7 22.2 2.3	3 11.5 37.5 3.4	5 19.2 50.0 5.7	6 23.1 42.9 6.8	2 7.7 66.7 2.3	2 7.7 66.7 2.3	2 7.7 40.0 2.3	26 29.5
SPECIAL	2 14.3 18.2 2.3	4 28.6 25.0 4.5	1 7.1 11.1 1.1	1 7.1 11.1 1.1	1 7.1 12.5 1.1	2 14.3 20.0 2.3	2 14.3 14.3 2.3	0 0.0 0.0 0.0	1 7.1 33.3 1.1	0 0.0 0.0 0.0	14 15.9
LAB SUB	4 19.0 36.4 4.5	5 23.8 31.3 5.7	3 14.3 33.3 3.4	1 4.8 11.1 1.1	3 14.3 37.5 3.4	1 4.8 10.0 1.1	2 9.5 14.3 2.3	1 4.8 33.3 1.1	0 0.0 0.0 0.0	1 4.8 20.0 1.1	21 23.9
L&M SUB	4 14.8 36.4 4.5	5 18.5 31.3 5.7	4 14.8 44.4 4.5	5 18.5 55.6 5.7	1 3.7 12.5 1.1	2 7.4 20.0 2.3	4 14.8 28.6 4.5	0 0.0 0.0 0.0	0 0.0 0.0 0.0	2 7.4 40.0 2.3	27 30.7
Column Total %	11 12.5	16 18.2	9 10.2	9 10.2	8 9.1	10 11.4	14 15.9	3 3.4	3 3.4	5 5.7	88 100.0

Refer to questions G_5_3 to G_5_6 of the original questionnaire for the original questions

N.B.   The values used to compile this table are constant 1992 values, derived by updating the original values with the Consumer Price Index. The original values applied to the first contract undertaken by the firm, and it was assumed that this first contract occurred in the same year that the firm started operating.

Table 5.28 - Monthly Turnover on Very First Contract

Count / Row % / Col % / Total %	LESS THAN R2 000	R2 001 TO R4 000	R4 001 TO R6 000	R6 001 TO R8 000	R8 001 TO R10 000	R10 001 TO R15 000	R15 001 TO R25 000	R25 001 TO R35 000	R35 001 TO R60 000	MORE THAN R60 000	Row Total
GENERAL Count	2	4	7	3	1	3	3	3	0	0	
Row %	7.7	15.4	26.9	11.5	3.8	11.5	11.5	11.5	0.0	0.0	
Col %	25.0	16.0	43.8	27.3	14.3	42.9	42.9	50.0	0.0	0.0	
Total %	2.3	4.5	8.0	3.4	1.1	3.4	3.4	3.4	0.0	0.0	
SPECIAL Count	2	3	1	2	2	0	0	3	0	1	
Row %	14.3	21.4	7.1	14.3	14.3	0.0	0.0	21.4	0.0	7.1	
Col %	25.0	12.0	6.3	18.2	28.6	0.0	0.0	50.0	0.0	100.0	
Total %	2.3	3.4	1.1	2.3	2.3	0.0	0.0	3.4	0.0	1.1	
LAB SUB Count	4	8	2	3	3	0	1	0	0	0	
Row %	19.0	38.1	9.5	14.3	14.3	0.0	4.8	0.0	0.0	0.0	
Col %	50.0	32.0	12.5	27.3	42.9	0.0	14.3	0.0	0.0	0.0	
Total %	4.5	9.1	2.3	3.4	3.4	0.0	1.1	0.0	0.0	0.0	
L&M SUB Count	0	10	6	3	1	4	3	0	0	0	
Row %	0.0	37.0	22.2	11.1	3.7	14.8	11.1	0.0	0.0	0.0	
Col %	0.0	40.0	37.5	27.3	14.3	57.1	42.9	0.0	0.0	0.0	
Total %	0.0	11.4	6.8	3.4	1.1	4.5	3.4	0.0	0.0	0.0	
Column Count	8	25	16	11	7	7	7	6	0	1	
Total %	9.1	28.4	18.2	12.5	8.0	8.0	8.0	6.8	0.0	1.1	

Refer to questions G_6_3 to G_6_6 of the original questionnaire for the original questions

N.B.   It was assumed that, in all cases, the most recent contract was undertaken in 1992. The data in the table are thus constant 1992 val

Table 5.29 - Monthly Turnover on Most Recent Contract

A comparison of these two tables shows that there was movement in the distribution of the monthly turnovers across the value categories. There was a decrease in the 'less than R2 000' category. This suggests that growth occurred among those who first undertook lower value contracts (and who are like be sub-contractors). Given that 1992 constant prices are being compared, a movement into a higher turnover value category probably indicates that the firm had successfully progressed to a more demandi kind of work (see section 2.4.4 regarding lateral and upward expansion). Indeed, this notion is supporte by the fact that labour-and-material sub-contractors accounted for 36.4% of those with monthly turnove under R2 000 on their first contract, but on their most recent contracts, this figure had reduced to 0.0% Notably, no movement occurred in the numbers of specialist sub-contractors and labour-only sub-contractors whose turnovers were less than R2 000.

Numbers decreased in all of the value categories exceeding R8 000, with one exception - the 'R25 001 R35 000' value category. The increase in this case was largely attributable to the increase in specialist s contractors with turnovers in this category. Other increases of significance included an increase in the 'R2 000 to R4 000' category and also in the 'R4 000 to R6 000' category. In these two categories combined, numbers increased from 28.4% to 46.6%.

From Table 5.29 it is seen that the majority of firms (55.7%) turned over less than R6 000 per month t they were executing their most recent contract.

(d) Profit: The data for questions G_5_3 (actual income from contract) and G_5_5_1 to G_5_5_4 (total costs) was used to determine the firms' profits for their first and most recent contracts. The declared profit (see question G_5_7) was deducted from the result of this calculation to determine whether or not the declared profit was accurate.

Count Column %	VERY FIRST CONTRACT	MOST RECENT CONTRACT
NUMBER WHO UNDER-CALCULATED PROFIT	13 14.8	15 17.0
NUMBER WHO OVER-CALCULATED PROFIT	24 27.3	30 34.1
NUMBER WHO CALCULATED PROFIT CORRECTLY	51 58.0	43 48.9
Column Total %	88 100.0	88 100.0

Refer to questions G-5-7 and G_6_7 of the questionnaire for the original questions

Table 5.30 - Comparison Between Accuracy of Profit Calculation for First and Most Recent Contracts

Table 5.30 reveals some disturbing results. It was expected that all respondents would be able to calculate their profit correctly, since they themselves provided the values for income and for costs. However, concerning their first contracts, only 58% calculated profit correctly, 27.3% thought they had made more profit than they actually did, while 14.8% thought they had made less profit than they actually did. The corresponding data for the most recent contracts showed that 48.9% calculated profit accurately, 34.1% thought they had made more profit than was actually the case and 17.0% thought they had made less profit than was actually the case.

The data for the first contract were generally provided many years after the contract had been completed. This raises doubts as to whether or not the data are accurate. Thus, it cannot be said with certainty that the respondents became less capable of calculating profit accurately since their first contracts, which is what the results in the above table suggest. Nevertheless, significant numbers of respondents could not calculate profit accurately from their own figures. These data tend to support those in section 2.5.1 where it was noted that SSCEs are generally poor at cost management and bookkeeping.

The actual amounts of profits and losses experienced on first and most recent contracts are reflected in Tables 5.31 and 5.32. Values for profit on first contracts were updated to 1992 constant prices using the Consumer Price Index, while values of profit on most recent contracts were assumed to be 1992 values.

Count Row % Col % Total %	LESS THAN MINUS R5 000	MINUS R4 999 TO MINUS R2 500	MINUS R2 499 TO ZERO	R1 TO R1 000	R1 001 TO R2 000	R2 001 TO R5 000	R5 001 TO R10 000	R10 001 TO R20 000	R20 001 TO R50 000	MORE THAN R50 000	Row Total
GENERAL	0	0	6	1	0	3	5	1	7	3	[illegible]
	0.0	0.0	23.1	3.8	0.0	11.5	19.2	3.8	26.9	11.5	
	0.0	0.0	46.2	7.7	0.0	20.0	45.5	20.0	46.7	75.0	
	0.0	0.0	6.8	1.1	0.0	3.4	5.7	1.1	8.0	3.4	
SPECIAL	2	0	2	4	1	3	0	0	2	0	[illegible]
	14.3	0.0	14.3	28.6	7.1	21.4	0.0	0.0	14.3	0.0	
	66.7	0.0	15.4	30.8	12.5	20.0	0.0	0.0	13.3	0.0	
	2.3	0.0	2.3	4.5	1.1	3.4	0.0	0.0	2.3	0.0	
LAB SUB	1	0	2	3	5	5	0	2	3	0	[illegible]
	4.8	0.0	9.5	14.3	23.8	23.8	0.0	9.5	14.3	0.0	
	33.3	0.0	15.4	23.1	62.5	33.3	0.0	40.0	20.0	0.0	
	1.1	0.0	2.3	3.4	5.7	5.7	0.0	2.3	3.4	0.0	
L&M SUB	0	1	3	5	2	4	6	2	3	1	[illegible]
	0.0	3.7	11.1	18.5	7.4	14.8	22.2	7.4	11.1	3.7	
	0.0	100.0	23.1	38.5	25.0	26.7	54.5	40.0	20.0	25.0	
	0.0	1.1	3.4	5.7	2.3	4.5	6.8	2.3	3.4	1.1	
Column Total %	3	1	13	13	8	15	11	5	15	4	[illegible]
	3.4	1.1	14.8	14.8	9.1	17.0	12.5	5.7	17.0	4.5	

N.B. The values used to compile this table are constant 1992 values, derived by updating the original values with the Consumer Price Inde[...] The original values applied to the first contract undertaken by the firm, and it was assumed that this first contract occurred in the sam[...] year that the firm started operating.

Table 5.31 - Amount of Profit on Very First Contract

Count Row % Col % Total %	LESS THAN MINUS R5 000	MINUS R4 999 TO MINUS R2 500	MINUS R2 499 TO ZERO	R1 TO R1 000	R1 001 TO R2 000	R2 001 TO R5 000	R5 001 TO R10 000	R10 001 TO R20 000	R20 001 TO R50 000	MORE THAN R50 000	Row Total
GENERAL	1	2	4	5	1	6	3	1	3	0	[illegible]
	3.8	7.7	15.4	19.2	3.8	23.1	11.5	3.8	11.5	0.0	
	50.0	100.0	44.4	38.5	7.7	31.6	18.8	12.5	75.0	0.0	
	1.1	2.3	4.5	5.7	1.1	6.8	3.4	1.1	3.4	0.0	
SPECIAL	1	0	1	0	3	0	5	2	0	2	[illegible]
	7.1	0.0	7.1	0.0	21.4	0.0	35.7	14.3	0.0	14.3	
	50.0	0.0	11.1	0.0	23.1	0.0	31.3	25.0	0.0	100.0	
	1.1	0.0	1.1	0.0	3.4	0.0	5.7	2.3	0.0	2.3	
LAB SUB	0	0	2	4	5	6	3	1	0	0	[illegible]
	0.0	0.0	9.5	19.0	23.8	28.6	14.3	4.8	0.0	0.0	
	0.0	0.0	22.2	30.8	38.5	31.6	18.8	12.5	0.0	0.0	
	0.0	0.0	2.3	4.5	5.7	6.8	3.4	1.1	0.0	0.0	
L&M SUB	0	0	2	4	4	7	5	4	1	0	[illegible]
	0.0	0.0	7.4	14.8	14.8	25.9	18.5	14.8	3.7	0.0	
	0.0	0.0	22.2	30.8	30.8	36.8	31.3	50.0	25.0	0.0	
	0.0	0.0	2.3	4.5	4.5	8.0	5.7	4.5	1.1	0.0	
Column Total %	2	2	9	13	13	19	16	8	4	2	[illegible]
	2.3	2.3	10.2	14.8	14.8	21.6	18.2	9.1	4.5	2.3	

N.B. It was assumed that, in all cases, the most recent contract was undertaken in 1992. Thus these data are constant 1992 values.

Table 5.32 - Amount of Profit on Most Recent Contract

The first fact that can be observed from a comparison between Tables 5.31 and 5.32 is that 19.3% of the respondents made losses on their first contracts and that this reduced slightly to 14.8% on most recent contracts. The second observation that should be made is that there were increases in the number of respondents who made profits in each of the value categories, except in the R1 to R1 000 (which remained static), and those over R20 000 (which both decreased).

The majority of firms that made a profit, made less than R5 000 per contract for both first and most recent contracts. Seventy-one (71) firms made profits on their first contracts and 50.7% of these firms made less than R5 000 profit. Seventy-five (75) firms made profits on their most recent contracts and 60.0% of these made less than R5 000 profit. The increase in the number of profit makers who made less than R5 000 indicates that the tendency was for firms to make less profit on their most recent contracts than was the case on their first contracts. If amount of profit made can be taken as an indicator of growth, then these data suggest that, generally, firms experienced negative growth.

Looking now at the trends for the various type of contractors, it is found that slightly more general contractors made losses on their most recent contracts than on their first contracts, and the losses tended to be greater. This group also tended to make lower profits on their most recent contracts. Similar trends can be seen in the data for labour-only sub-contractors.

In the case of specialist contractors, however, fewer made a loss on their most recent contracts than on their first contracts and those who made a profit, generally made higher profits. Similarly, fewer labour-and-material sub-contractors made losses on their most recent contracts, but where profits were made, these tended to be lower than on their first contracts.

Tables 5.33 and 5.34 reflect profit percentages, calculated by expressing profit over cost. It must be noted that these profit percentages appear to be unusually high. This is probably due to the fact that contract amounts tended to be low. A comparison of these two tables confirms the suggestion that negative growth had occurred. This is seen in the data which show that the majority (38, or 55.1%, of the 69 firms) who made profits on their first contracts, made more than 50% profit. However, the majority (40, or 54.8%, of the 73 firms) who made profits on their most recent contracts, made more than 40% profit.

5.7.6 How Success Is Measured

Table 5.35, contains the responses of firm owners when asked how they measure success in their businesses. The most commonly cited criteria were: profit; approaches from clients; client satisfaction; and having a lot of current work.

Count Row % Col % Total %	LOSS	BREAK EVEN	1% TO 10%	11% TO 15%	16% TO 20%	20% TO 30%	31% TO 40%	41% TO 50%	51% TO 75%	MORE THAN 75%	Row Total
GENERAL	5 20.0 41.7 5.8	1 4.0 20.0 1.1	2 8.0 40.0 2.3	1 4.0 50.0 1.1	2 8.0 50.0 2.3	1 4.0 11.1 1.1	1 4.0 20.0 1.1	2 8.0 33.3 2.3	6 24.0 35.3 7.0	4 16.0 19.0 4.7	[illegible]
SPECIAL	3 23.1 25.0 3.5	1 7.7 20.0 1.1	1 7.7 20.0 1.1	0 0.0 0.0 0.0	0 0.0 0.0 0.0	2 15.4 22.2 2.3	2 15.4 40.0 2.3	0 0.0 0.0 0.0	4 30.8 23.5 4.7	0 0.0 0.0 0.0	[illegible]
LAB SUB	2 9.5 16.7 2.3	1 4.8 20.0 1.1	1 4.8 20.0 1.1	0 0.0 0.0 0.0	0 0.0 0.0 0.0	5 23.8 55.6 5.8	0 0.0 0.0 0.0	0 0.0 0.0 0.0	2 9.5 11.8 2.3	10 47.6 47.6 11.6	[illegible]
L&M SUB	2 7.4 16.7 2.3	2 7.4 40.0 2.3	1 3.7 20.0 1.1	1 3.7 50.0 1.1	2 7.4 50.0 2.3	1 3.7 11.1 1.1	2 7.4 40.0 2.3	4 14.8 66.7 4.7	5 18.5 29.4 5.8	7 25.9 33.3 8.1	[illegible]
Column Total %	12 14.0	5 5.8	5 5.8	2 2.3	4 4.7	9 10.5	5 5.8	6 7.0	17 19.8	21 24.4	[illegible]

N.B. Profit percentage = profit ÷ cost.

Table 5.33 - Profit Percentage on Very First Contract

Count Row % Col % Total %	LOSS	BREAK EVEN	1% TO 10%	11% TO 15%	16% TO 20%	20% TO 30%	31% TO 40%	41% TO 50%	51% TO 75%	MORE THAN 75%	Row Total
GENERAL	6 23.1 66.7 7.0	1 3.8 25.0 1.2	3 11.5 50.0 3.5	1 3.8 33.3 1.2	2 7.7 28.6 2.3	2 7.7 18.2 2.3	3 11.5 50.0 3.5	4 15.4 50.0 4.7	1 3.8 6.7 1.2	3 11.5 17.6 3.5	[illegible]
SPECIAL	1 7.1 11.1 1.2	1 7.1 25.0 1.2	1 7.1 16.7 1.2	0 0.0 0.0 0.0	1 7.1 14.3 1.2	2 14.3 18.2 2.3	0 0.0 0.0 0.0	0 0.0 0.0 0.0	6 42.9 40.0 7.0	2 14.3 11.8 2.3	[illegible]
LAB SUB	2 10.5 22.2 2.3	0 0.0 0.0 0.0	0 0.0 0.0 0.0	1 5.3 33.3 1.2	0 0.0 0.0 0.0	2 10.5 18.2 2.3	1 5.3 0.0 0.0	0 0.0 0.0 0.0	7 36.8 46.7 8.1	6 31.6 35.3 7.0	[illegible]
L&M SUB	0 0.0 0.0 0.0	2 7.4 50.0 2.3	2 7.4 33.3 2.3	1 3.7 33.3 1.2	4 14.8 57.1 4.7	5 18.5 45.5 5.8	2 7.4 40.0 2.3	4 14.8 50.0 4.7	1 3.7 6.7 1.2	6 22.2 35.3 7.0	[illegible]
Column Total %	9 10.5	4 4.7	6 7.0	3 3.5	7 8.1	11 12.8	6 7.0	8 9.3	15 17.4	17 19.8	[illegible]

N.B. Profit percentage = profit ÷ cost.

Table 5.34 - Profit Percentage on Most Recent Contract

The column percentages in Table 5.35 exceed 100.0%, indicating that some (22) respondents gave two answers to the question. Profit was the most common measure of success. This was expected, but, as was seen in Table 5.30, more than half of the respondents were unable to accurately calculate their profit on their most recent contracts. This suggests that many firms might think that they are successful, when in reality, they are particularly unsuccessful. 'Frequency of approaches by clients', and 'degree of client satisfaction' were, respectively, the second and third most common measures of success.

Count Column %	NUMBER	Column %
Completing jobs	3	3.1
Profit	25	26.0
Client satisfaction	11	11.5
Having many current jobs	11	11.5
Being approached by many clients	19	19.8
Continued invitations to sub-contract for main contractor	1	1.0
Having enough current jobs to survive	1	1.0
Regular inflow of work	1	1.0
Increasing capital	6	6.3
Getting more future work	10	10.4
Getting jobs of high contract value	2	2.1
Growth in staff numbers	1	1.0
Expansion of the firm	1	1.0
Degree to which jobs can be sub-contracted to others	1	1.0
Completing jobs before due date	3	3.1
Doing quality work	2	2.1
Following up on debtors	1	1.0
Getting prestigious buildings	2	2.1
Getting large buildings	1	1.0
Getting paid the agreed amount	1	1.0
No response	14	14.6
Column Total %	117	121.9

Refer to question G_9 of the questionnaire for the original question

N.B.   In 22 cases, the respondent cited two measures of success. For this reason the column percentages in the above table (which are percentages of the survey size of 96 respondents, exceed 100%.

Table 5.35 - How Firms Determine Success

It is interesting to note that 'increasing capital' was not commonly cited as a measure of success. This suggests that profits are not ploughed back into the business, but are either consumed, or serve as bridging finance for future contracts.

5.8   FUTURE OBJECTIVES OF THE FIRMS

In this section the objectives of the firms are described. Included are aspects such as: what the firm plan do; if these plans include expansion, into which markets; steps taken towards fulfilling objectives; and desired shifts in the type of contracting currently undertaken.

5.8.1   Plans for the Future

Notably, no firm owners intended leaving the building industry. This is seen from the data on page 182 Appendix H. From this source, it can also be seen that the majority (86.5%) of the respondents intended expand their businesses. Of this sample, 25.3% intended obtaining most of their work from private families, 32.5% from contractors, 30.1% from provincial or government agencies, and 12.0% from a combination of these types of employer.

5.8.2   Steps Taken Towards Fulfilling Objectives

The following table, Table 5.36, contains the responses concerning what action had been taken towards meeting future objectives.

Count Col %	NUMBER
Tried to diversify	18 18.75
Advertised	45 46.9
Invested in additional capital	29 30.2
Approached possible principal constructors	25 26.04
Invested / saving	49 51.04
Column Total %	166 172.9

Refer to question H_3 of the questionnaire for the original question

N.B.   The question allowed for multiple responses. For this reason the column
percentages in the above table (which are percentages of the total survey
size of 96), exceed 100%.

Table 5.36 - Steps Taken Towards Achieving Objectives

The majority of respondents had started saving towards the fulfilment of their objectives. Other common steps taken included advertising the business to attract more clients and investment in additional capital. It should also be noted that the question attracted multiple responses, with many respondents indicating that they had taken more than one course of action.

5.8.3 Desired Changes in Type of Activity Performed in the Building Industry

Table 5.37, below, presents the responses of two sub-groupings of contractors regarding what type of contracting work they would prefer if they could change from their current type of work. These sub-groupings consisted of the 12 general contractors doing mainly alterations and additions work and all types of sub-contractors.

Count Col %	GENERAL CONTRACTORS DOING MAINLY ALTERATIONS & ADDITIONS WORK	ALL TYPES OF SUB-CONTRACTOR
Keep on doing alterations and additions work	4 33.3	5 7.6
Become a labour-only sub-contractor	0 0.0	6 9.1
Become a plumber	0 0.0	0 0.0
Become an electrician	1 8.3	11 16.7
Become a general contractor doing mainly new work	5 41.7	33 50.0
None of these	2 16.7	11 16.7
Column Total %	12 100.0	66 100.0

Refer to questions H_4 and H_5 of the questionnaire for the original questions

Table 5.37 - Desired Changes in Type of Contracting

Some important results emerged from this table. Only 33.3% of the general contractors wanted to continue doing alterations and additions work. The majority of those who wished to change, would have preferred to remain as general contractors, but doing mainly new work. Similar desires were expressed by the sub-contracting group, of whom 50.0% indicated that they would prefer to be general contractors doing mainly new work.

The importance of these findings lies in two inferences. Firstly, it appears that general contractors gain entry to the industry by initially taking on alterations and additions jobs, but only because they cannot obtain any other kind of work. Secondly, sub-contractors generally appear to be unhappy with their status and would prefer to contract directly with their clients, rather than with another contractor. These data confirm Merrifield's (1992:61) observation that sub-contractors aspire to general contracting (see section 2.4.2)

5.9   MANAGEMENT/PRODUCTION

In this section the data pertaining to the management of the business and its site operations are analysed. Aspects covered include: sources of equipment and technical support; banking practices; tendering metho reasons for not completing contracts; working capital; and capacity to do contracts simultaneously.

5.9.1   Sources of Equipment and Technical Support

It was found that 82.3% of the respondents own their equipment, 49.0% hire it, 3.1% lease it, and 2.1% borrow it from friends (pgs.191-194/AppH;Q1_1). The total of these percentages exceed 100.0% because the question attracted multiple responses. Notably, leasing was found to be an uncommon method of obtaining equipment. This is related to the previous finding that access to finance was the most serious problem experienced by firms at start-up (see section 5.6.3). Failure to qualify for finance from a formal institution would almost certainly mean that the applicant would fail to secure a lease contract through su an institution.

Count \ Col %	NUMBER	Column %
Joint venture partner	22	22.9
Equipment suppliers	28	29.2
Government, SBDC, CSIR	15	15.6
Industry association	15	15.6
General contractor who employed you	20	20.8
Materials suppliers	28	29.2
Materials producers/manufacturers	18	18.8
Non-government organisations	4	4.2
Private consultants	17	17.7
Column Total %	96	100.0

Refer to question 1_2 of the questionnaire for the original question

Table 5.38 - Sources of Technical Support

Short-term hiring (by the hour, or day) was the most common source of equipment not already owned by the respondents. These data support the literature survey finding that SSCEs tend to use labour intensive methods (see section 2.5.2).

Table 5.38 presents data regarding sources of technical support. The three major sources of this type of support were: materials suppliers (29.2%); equipment suppliers (29.2%); and joint venture partners (22.9%). General contractors can also be seen to have provided support to their sub-contractors. An important finding was that NGOs and contractors' associations were relatively minor sources of technical support - this tends to support the arguments of Chapter 4.

5.9.2   Banking and Financial Record Keeping

The data in Table 5.39 reflect details of whether or not bank accounts were run by the respondents, and whether or not business and personal transactions were kept separate.

Count Row %	SEPARATE PERSONAL AND BUSINESS BANK ACCOUNTS	ONE BANK ACCOUNT FOR BOTH PERSONAL AND BUSINESS	NO BANK ACCOUNT	NO RESPONSE	Row Total %
ALL	54 56.3	33 34.4	5 5.2	4 4.2	96 100.0

Refer to question 1_3 of the questionnaire for the original question

Table 5.39 - Type of Bank Account Operated

A high percentage (39.6%) of the respondents reported that they either ran a combined account for business and personal transactions, or that they did not run a bank account at all. In addition, it was found that 67.7% of the respondents kept accurate financial records of their business transactions (pg.205/AppH;Q-1_4). These two findings are clearly related to the previous result which showed lack of access to finance to be the most common serious problem at start-up (see section 5.6.3). Notably the incidence of financial record-keeping in the sample for this survey was higher than for Krafchik's (1990:228) sub-contractors (see section 2.5.1). However, it must be noted that the findings of Table 5.30 suggest that, while respondents were found to keep records, these cannot be described as accurate records.

5.9.3   Tendering

A significant number (37.5%) of respondents reported that they had experienced problems calculating tender amounts, while 53.1% had not had this problem (pg.205/AppH;Q-1_5). This latter percentage is deceptively high, and is probably not a true reflection of the situation. This is suggested by the fact that negotiation, rather than competitive tender, was found to be the most common method of obtaining work - and that take-it-or-leave-it offers from clients were not uncommon (see section 5.7.3). It can be said that

these methods of acquiring work would not necessarily involve sophisticated costing skills and in the ligh
this, it is not surprising that such a high percentage of respondents did not regard the calculation of a tend
amount as problematic.

Count Row %	CALCULATE RATE PER SQUARE METRE	CALCULATE QUANTITIES AND COSTS OF EVERYTHING	FIND OUT COMPETITORS COST AND CHARGE LESS	NO RESPONSE	Row Total %
ALL	34 35.4	41 42.7	12 12.5	9 9.4	10

Refer to question I_6 of the questionnaire for the original question

Table 5.40 - How Tender Amounts are Established

In Table 5.40, details are given of how the respondents arrive at their tender amounts. Notably, only 42.
reported that they build the tender figure up by quantifying and costing all items of material, labour,
equipment, *etc.*. A significant number (35.4%) of the respondents noted that they use a square metre rat
to establish the tender amount, while 12.5% reported that they do not calculate the amount, but simply
undercut their competitors.

The use of either of these latter two methods of compiling tender amounts (which together were employe
by 47.9% of the respondents), would leave a contractor completely uninformed as to what his costs woul
be for labour, materials, equipment, *etc.*. And, considering that a detailed breakdown of costs is an
essential tool for planning the financial management of the contract, financial planning and control would
difficult, if not impossible.

These data tend to support the literature survey finding that estimating and tendering are internal constrai
of SSCEs (see section 2.5.1)

5.9.4 Working Capital

As noted in the previous section, the majority of respondents compile their tender amounts in such a
manner that they would have insufficient cost data to plan the financial management of the contract.
However, 70.8% reported that they usually calculated working capital requirements (pg.214/AppH;Q-I_8
This suggests that financial planning occurs while the contract is in progress, rather than prior to its
commencement.

The following table (Table 5.41) details the frequency of working capital calculations.

Count Row %	1 WEEK IN ADVANCE	1 MONTH IN ADVANCE	3 MONTHS IN ADVANCE	1 YEAR IN ADVANCE	Row Total %
ALL	41 59.4	25 36.2	2 2.9	1 1.4	69 100.0

Refer to question 1_8_1 of the questionnaire for the original question

Table 5.41 - Frequency of Working Capital Calculation

These data confirm the suggestion that financial planning mainly occurs while the contract is being executed. This can be seen from the fact that the majority (59.4%) of those who do calculate working capital requirements, do so one week in advance. Indeed, only 4.3% of this sample showed any tendency towards medium or long-term financial planning. These data concur with the identification in the literature that the inability to plan projects is an internal constraint faced by SSCEs (see section 2.5.1 e).

5.9.5 Premature Departure from Site

The findings discussed in the previous two sections suggest that many respondents might have entered into contracts and later discovered that they were bound to make losses. Questions were included which required respondents to indicate whether or not they had ever left a site before completing the contract, and if so, why. Many might have been unwilling to admit to this and it is consequently likely that the following percentages are understated. Notwithstanding this, 40.6% of the respondents noted that they had left sites prematurely, while 51.0% had never done so (pg.208/AppH;Q-1_7).

Count	Col %	NUMBER	Column %
Not paid	27	69.2
Ran out of finance	9	23.1
Forced out because of threats	6	15.4
General level of violence in the area	19	48.7
Materials were being stolen	1	2.6
Column	Total %	70	179.5

Refer to question I_7 of the questionnaire for the original question

N.B.   The question allowed for multiple responses. For this reason, column percentages (which are percentages of the total of 39 respondents who had left sites prematurely), exceed 100.0%.

Table 5.42 - Reasons for Leaving Sites Prematurely

In Table 5.42, above, reasons are given regarding why sites had been abandoned. The most common reason given was 'not paid'. It is not clear from the responses whether this was because of poor quality work, or whether clients could not afford to pay. In any event, it is more likely that clients believed they had good

reason to withhold payment than that they refused to pay for work with which they were satisfied. It is
likely, though, that 'not paid' and 'ran out of finance' were interpreted as meaning the same thing, that
lack of finance. These data go a long way to explaining the generally poor reputation of black SSCEs (s
section 2.5.3.

The second most common response was that contracts were terminated prematurely because of the high
level of violence in the area. It is important to note that the markets in which most of the respondents
usually operate are somewhat different from those in which most of their white counterparts work. Failu
to complete a contract is possibly the most unacceptable action a contractor could take, and would impac
extremely negatively on his reputation. Thus, it is clear that violence and unrest, which are products of
political turmoil occurring in South Africa at present, cause contractors to take actions which contribute
significantly to their poor reputations. This finding was expected since high levels of violence in townsh
have reportedly caused support institutions to fail (see section 4.1).

5.9.6   Capacity to Do Contracts Simultaneously

The following table (Table 5.43) contains data indicating how many contracts could be undertaken
simultaneously by the firms. The purpose of the question was to establish whether or not the managemen
structure and techniques of the firms were of such a nature that they could permit or facilitate expansion

Count Row %	NO RESPONSE	CAN ONLY DO 1 AT A TIME	2	3	4 TO 10	11 TO 15	15 TO 23	Row Tot
ALL	8 8.3	23 24.0	29 30.2	17 17.7	16 16.7	1 1.0	2 2.1	

Refer to question I_9 of the questionnaire for the original question

Table 5.43 - Number of Contracts Capable of Being Undertaken Simultaneously

From Table 5.43 it is clear that the majority (54.2%) of firms could not undertake more than two contra
simultaneously. Indeed, 24.0% could only do one job at a time. The importance of these findings lies in
their relationship with the previous finding that 86.5% of the respondents reported that they intended
expanding their businesses (see section 5.8.1). For the majority, expansion would thus have to be vertica
since lateral expansion would require that more contracts are done simultaneously (see section 2.4.4).
However, vertical expansion would require that bigger projects be undertaken, and it can be seen from t
data in section 5.7 generally, that this is unlikely to be accomplished without a substantial degree of
support. These data thus tend to further support the contention that the capacity SSCEs is limited (see
Table 5.25; Merrifield, 1992 and; section 2.5.2 f)

5.10   CAPITAL AND FINANCING

This section describes: the sources of capital and finance of the firms; problems experienced by firms regarding obtaining loans; legal judgements and their effect on the firm; and growth in capital over the life of the firm.

5.10.1 Sources of Finance

In Table 5.44, below, the data indicates the firms' major sources of finance - at start-up and in 1992. The question attracted multiple responses. Totals therefore exceed 100.0%, indicating that many respondents obtained finance from more than one source.

Count Col %	SOURCES OF FINANCE AT START-UP	SOURCES OF FINANCE IN 1992
Own savings	68 70.8	80 83.3
Family and friends	43 44.8	19 19.8
Partners	21 21.9	19 19.8
Informal moneylenders	8 8.3	3 3.1
Formal commercial bank	9 9.4	7 7.3
Government agencies (e.g. SBDC)	6 6.3	5 5.2
Welfare, churches and other donors	1 1.0	1 1.0
Credit from main contractor	4 4.2	6 6.3
Credit from material and equipment suppliers	13 13.5	7 7.3
Advance payments from employer	25 26.0	18 18.8
Development Corporation	0 0.0	0 0.0
Column Total %	198 206.3	165 171.9

Refer to question J_1 of the questionnaire for the original question

N.B.   The question allowed for multiple responses. For this reason the column percentages in the above table (which are percentages of the total survey size of 96) exceed 100%.

Table 5.44 - Sources of Finance at Start-up and in 1992

The first important observation which must be made is that the most common source of finance, both at start-up and in 1992, was the respondents' own savings. Indeed, as can clearly be seen from the table, th became a more common source of finance as firms got older. This finding confirms the suggestion (noted section 5.7.6) that profits are not invested in the business, but rather are used as a source of bridging finance for future contracts.

Family and friends were another major source of finance, both at start-up and in 1992. There was, however, a sharp decrease in the numbers of firms that obtained money from this source as they became more established. Thus, family and friends play a major role in helping the firm to become established, whereafter there is a much greater reliance on savings as a source of finance.

Advance payments from employers were a relatively important source of finance throughout the life of th firms, although there tended to be a greater reliance on savings as firms grew older.

Another important source of finance, at start-up and in 1992, was the respondents' partners. This suggest that partnerships may be entered into for the purpose of securing a source of finance, rather than with th aim of broadening the firms' skills base. This speculation is based on the finding (in section 5.2.4) that 24.0% of the respondents reported that they had business partners and 21.9% cited 'partners' as a source initial finance, while 19.8% still obtained finance from their partners in 1992 (see Table 5.44).

Notably, formal commercial banks were an uncommon source of finance, as were government agencies, material or equipment suppliers, and informal moneylenders. All of these sources were more common initial, rather than current (1992) sources.

The findings in this section confirm that, as noted previously and in the literature (see section 2.5.2), acc to finance is a problem for SSCEs. In addition, the data tend to confirm the argument in Chapter 4 that t SBDC is essentially beyond the reach of most black SSCEs.

5.10.2 Loan Collateral Problems

The results reflected in Table 5.45, below, indicate whether or not the respondents experienced problems meeting collateral requirements when applying for loans.

The finding that the majority (51.0%) of the respondents answered 'not applicable' (see Table 5.45) must be considered together with the results in Table 5.44, where it was clearly shown that formal commercial banks were an insignificant source of finance. It is therefore not surprising that this high percentage indicated that the question did not apply to them.

Count Row %	NOT APPLICABLE (DO NOT APPLY FOR LOANS) OR NO RESPONSE	DO EXPERIENCE PROBLEMS MEETING SECURITY / COLLATERAL REQUIREMENTS WHEN APPLYING FOR LOANS	DO NOT EXPERIENCE PROBLEMS MEETING SECURITY / COLLATERAL REQUIREMENTS WHEN APPLYING FOR LOANS	Row Total %
ALL	49 51.0	33 34.4	14 14.6	96 100.0

Refer to question J_2 of the questionnaire for the original question

Table 5.45 - Numbers Who Experienced Collateral Problems

As can be seen in Table 5.45, only 47 (49.0%) of the total number of respondents had made loan applications which required collateral. Thirty-three (33) of these 47 respondents, or 70.2%, reported that they had experienced problems meeting collateral requirements. In Table 5.46 the responses of this group regarding what the problem actually was are presented. The question permitted multiple responses and the fact that 72 responses were recorded, therefore indicates that many respondents had experienced problems for more than one reason.

Count Col %	NUMBER
Do not own property	17 51.5
Do not have sufficient capital	22 66.7
Do not have contracts to code	9 27.3
Financial institution are inaccessible	9 27.3
Not been in business long enough	15 45.5
Column Total %	33 218.2

Refer to question J_2_1 of the questionnaire for the original question

N.B.   The question allowed for multiple responses. For this reason the column percentages in the above table (which are percentages of the total sample size of 33), exceed 100%.

Table 5.46 - Problems in Meeting Collateral Requirements

From Table 5.46 it is seen that the most common problems regarding meeting collateral requirements were: lack of capital; non-ownership of property; and lack of business track record. These data confirm the ILO findings that SSCEs struggle to meet collateral requirements, but the data also confirm the reasons for this as noted in the local literature - insufficiency of capital and lack of track record (see section 2.5.2).

Included in the questionnaire was a question aimed only at those who had obtained loans, which sought to

establish whether or not they had found loan repayment or interest rate fluctuations to be problematic.
Fourteen (14) of the sample of 47, or 29.8%, had experienced such problems, while 68.1% had not
(pg.246/AppH;Q-J_3). This suggests that the lending institutions had made loans to those who were
unlikely to default on repayment and that, in the majority of cases, their assessment of ability to repay w
correct.

5.10.3 Legal Judgements

Only 6 of the respondents indicated that legal judgements had been awarded against them and, of these,
only 1 indicated that the effect of this had been that he was unable to secure a loan (pgs.247-248/AppH;
J_4 & J_4_1). It must be noted that it would be potentially damaging, or at least embarrassing, to admit
this. These results are thus unlikely to be a true reflection of the real situation.

5.10.4 Money and Investments

Tables 5.47 and 5.48 give details of the firm's cash and capital investments at start-up, and in 1992. Va
for the year of start-up were converted to constant 1992 values (using the Consumer Price Index). The
original question required the respondent to indicate the values, at start-up and at the time of the intervi
of: cash; hand tools; transport (vehicles); electrically operated tools; petrol/diesel operated tools, *etc.*.

A comparison of the column totals in Tables 5.47 and 5.48, reveals that there was a general tendency fc
the total value of cash and capital investments to have increased between start-up and 1992. At start-up
largest single grouping of responses (25.3%) fell in the 'less than R5 000' category.

By 1992, however, the largest group of responses (16.1%) fell in the 'R5 000 to R10 000' category. Ar
analysis of the data by type of contractor reveals the following:

The majority (57.7%) of general contractors had cash and investments of less than R30 000 at start-up a
in 1992. Thus, apart from minor shifts within this value category, there was little of no growth in capita
over the life of the firm.

Notably, specialist sub-contractors reported generally higher values than the other types of contractors. 1
majority (69.3%) of specialist sub-contractors recorded values for cash and investments of less than
R50 000 at start-up. But by 1992, this percentage had reduced to 61.6% - indicating that some capital
growth had occurred.

Count Row % Col % Total %	LESS THAN R5 000	R5 001 TO R10 000	R10 001 TO R20 000	R20 001 TO R30 000	R30 001 TO R40 000	R40 001 TO R50 000	R50 001 TO R60 000	R60 001 TO R70 000	R70 001 TO R80 000	MORE THAN R80 000	Row Total %
GENERAL	2 7.7 9.1 2.3	3 11.5 25.0 3.4	6 23.1 50.0 6.9	4 15.4 66.7 4.6	3 11.5 27.3 3.4	3 11.5 37.5 3.4	0 0.0 0.0 0.0	0 0.0 0.0 0.0	1 3.8 50.0 1.1	4 15.4 66.7 4.6	26 29.9
SPECIAL	3 23.1 13.6 3.4	0 0.0 0.0 0.0	1 7.7 8.3 1.1	1 7.7 16.7 1.1	1 7.7 9.1 1.1	3 23.1 37.5 3.4	1 7.7 20.0 1.1	2 15.4 66.7 2.3	0 0.0 0.0 0.0	1 7.7 16.7 1.1	13 14.9
LAB SUB	12 57.1 54.5 13.8	3 14.3 25.0 3.4	3 14.3 25.0 3.4	0 0.0 0.0 0.0	1 4.8 9.1 1.1	0 0.0 0.0 0.0	1 4.8 20.0 1.1	1 4.8 33.3 1.1	0 0.0 0.0 0.0	0 0.0 0.0 0.0	21 24.1
L&M SUB	5 18.5 22.7 5.7	6 22.2 50.0 6.9	2 7.4 16.7 2.3	1 3.7 16.7 1.1	6 22.2 54.5 6.9	2 7.4 25.0 2.3	3 11.1 60.0 3.4	0 0.0 0.0 0.0	1 3.7 50.0 1.1	1 3.7 16.7 1.1	27 31.0
Column Total %	22 25.3	12 13.8	12 13.8	6 6.9	11 12.6	8 9.2	5 5.7	3 3.4	2 2.3	6 6.9	87 100.0

N.B. The values used to compile this table are constant 1992 values, derived by updating the original values with the Consumer Price Index. The original values applied to the first contract undertaken by the firm, and it was assumed that this first contract occurred in the same year that the firm started operating.

Table 5.47 - Value of Cash and Investments at Start-up

Count Row % Col % Total %	LESS THAN R5 000	R5 001 TO R10 000	R10 001 TO R20 000	R20 001 TO R30 000	R30 001 TO R40 000	R40 001 TO R50 000	R50 001 TO R60 000	R60 001 TO R70 000	R70 001 TO R80 000	MORE THAN R80 000	Row Total %
GENERAL	3 11.5 23.1 3.4	2 7.7 14.3 2.3	5 19.2 41.7 5.7	5 19.2 45.5 5.7	3 11.5 42.9 3.4	1 3.8 12.5 1.1	1 3.8 16.7 1.1	3 11.5 50.0 3.4	1 3.8 33.3 1.1	2 7.7 28.6 2.3	26 29.9
SPECIAL	2 15.4 15.4 2.3	0 0.0 0.0 0.0	1 7.7 8.3 1.1	2 15.4 18.2 2.3	2 15.4 28.6 2.3	1 7.7 12.5 1.1	1 7.7 16.7 1.1	0 0.0 0.0 0.0	1 7.7 33.3 1.1	3 23.1 42.9 3.4	13 14.9
LAB SUB	6 28.6 46.2 6.9	6 28.6 42.9 6.9	5 23.8 41.7 5.7	1 4.8 9.1 1.1	0 0.0 0.0 0.0	1 4.8 12.5 1.1	1 4.8 16.7 1.1	0 0.0 0.0 0.0	0 0.0 0.0 0.0	1 4.8 14.3 1.1	21 24.1
L&M SUB	2 7.4 15.4 2.3	6 22.2 42.9 6.9	1 3.7 8.3 1.1	3 11.1 27.3 3.4	2 7.4 28.6 2.3	5 18.5 62.5 5.7	3 11.1 50.0 3.4	3 11.1 50.0 3.4	1 3.7 33.3 1.1	1 3.7 14.3 1.1	27 31.0
Column Total %	13 14.9	14 16.1	12 13.8	11 12.6	7 8.0	8 9.2	6 6.9	6 6.9	3 3.4	7 8.0	87 100.0

N.B. Constant 1992 values

Table 5.48 - Value of Cash and Investments in 1992

Refer to question J_5 of the questionnaire for the original question relating to both tables 5.47 and 5.48.

Labour-only sub-contractors recorded much lower capital values than the other types of contractors, with 57.1% of them recording start-up capital values of less than R5 000. There was, however, some growth capital over the life of these firms, evidenced by the fact that in 1992, the same percentage (57.1%) recorded capital values of less than R10 000. The labour-and-material sub-contracting group recorded similar results. Capital growth was evidenced by the fact that 51.8% reported the start-up value of capital be less than R30 000, while the same percentage (51.8%) recorded 1992 capital values of less than R40 000.

In section 2.44 it was noted that some firms grow only until the owner has an acceptable personal income. The data in this section suggest that this is true for general contractors. The other groups did show growth, albeit slight, suggesting that sub-contracting is a growth area, or 'on-ramp' to general contracting (Rolfe 1990).

5.11   LABOUR

In this section details are given of the firms' employees. Aspects covered include: full-time and part-time employee numbers at start-up and in 1992; numbers of skilled employees; extent to which casual labour employed; wages; and extent to which general contractors employ sub-contractors.

5.11.1 Employee Numbers and Types of Employees

The questions on employee numbers required the respondent to give numbers of employees, specifying the types of employee and whether they were full-time, part time, casual, or family members.

For the purposes of this analysis, full-time employees are taken to include family members, and part-time employees are taken to include casuals. The following four tables (Tables 5.49 to 5.52) reflect numbers of employees for the various types of firm.

(a) Full-time employees: Tables 5.49 and 5.50 contain the data pertaining to full-time employees. A significant number of respondents reported having had no full-time employees, both at start-up (28.4%) and in 1992 (26.0%). There was a significant reduction, over the life of the firms, in the number of respondents who had employed only one skilled person - from 13.7% at start-up to 3.1% in 1992. Little movement occurred in the '2 to 5' and '6 to 10' categories, but in 1992 double the number of respondents employed 11 to 15 people than had been the case at start-up.

Count Row % Col % Total %	NONE	1	2 TO 5	6 TO 10	11 TO 15	16 TO 20	21 TO 30	31 TO 40	41 TO 50	MORE THAN 50	Row Total %
GENERAL	4	2	7	8	4	3	0	1	0	0	29
	13.8	6.9	24.1	27.6	13.8	10.3	0.0	3.4	0.0	0.0	30.5
	14.8	15.4	29.2	38.1	80.0	75.0	0.0	100.0	0.0	0.0	
	4.2	2.1	7.4	8.4	4.2	3.2	0.0	1.1	0.0	0.0	
SPECIAL	2	1	6	2	1	1	0	0	0	0	13
	15.4	7.7	46.2	15.4	7.7	7.7	0.0	0.0	0.0	0.0	13.7
	7.4	7.7	25.0	9.5	20.0	25.0	0.0	0.0	0.0	0.0	
	2.1	1.1	6.3	2.1	1.1	1.1	0.0	0.0	0.0	0.0	
LAB SUB	13	4	5	1	0	0	0	0	0	0	23
	56.5	17.4	21.7	4.3	0.0	0.0	0.0	0.0	0.0	0.0	24.2
	48.1	30.8	20.8	4.8	0.0	0.0	0.0	0.0	0.0	0.0	
	13.7	4.2	5.3	1.1	0.0	0.0	0.0	0.0	0.0	0.0	
L&M SUB	8	6	6	10	0	0	0	0	0	0	30
	26.7	20.0	20.0	33.3	0.0	0.0	0.0	0.0	0.0	0.0	31.6
	29.6	46.2	25.0	47.6	0.0	0.0	0.0	0.0	0.0	0.0	
	8.4	6.3	6.3	10.5	0.0	0.0	0.0	0.0	0.0	0.0	
Column Total %	27	13	24	21	5	4	0	1	0	0	95
	28.4	13.7	25.3	22.1	5.3	4.2	0.0	1.1	0.0	0.0	100.0

Refer to question K_1 of the questionnaire for the original question

Table 5.49 - Number of Full-time Employees at Start-up

Count Row % Col % Total %	NONE	1	2 TO 5	6 TO 10	11 TO 15	16 TO 20	21 TO 30	31 TO 40	41 TO 50	MORE THAN 50	Row Total %
GENERAL	4	1	8	4	7	2	1	0	2	0	29
	13.8	3.4	27.6	13.8	24.1	6.9	3.4	0.0	6.9	0.0	30.2
	16.0	33.3	34.8	16.0	70.0	66.7	33.3	0.0	100.0	0.0	
	4.2	1.0	8.3	4.2	7.3	2.1	1.0	0.0	2.1	0.0	
SPECIAL	1	0	3	6	0	1	1	1	0	1	14
	7.1	0.0	21.4	42.9	0.0	7.1	7.1	7.1	0.0	7.1	14.6
	4.0	0.0	13.0	24.0	0.0	33.3	33.3	100.0	0.0	100.0	
	1.0	0.0	3.1	6.3	0.0	1.0	1.0	1.1	0.0	1.1	
LAB SUB	12	1	4	5	0	0	1	0	0	0	23
	52.2	4.3	17.4	21.7	0.0	0.0	4.3	0.0	0.0	0.0	24.0
	48.0	33.3	17.4	20.0	0.0	0.0	33.3	0.0	0.0	0.0	
	12.5	1.0	4.2	5.2	0.0	0.0	1.0	0.0	0.0	0.0	
L&M SUB	8	1	8	10	3	0	0	0	0	0	30
	26.7	3.3	26.7	33.3	10.0	0.0	0.0	0.0	0.0	0.0	31.3
	32.0	33.3	34.8	40.0	30.0	0.0	0.0	0.0	0.0	0.0	
	8.3	1.0	8.3	10.4	3.1	0.0	0.0	0.0	0.0	0.0	
Column Total %	25	3	23	25	10	3	3	1	2	1	96
	26.0	3.1	24.0	26.0	10.4	3.1	3.1	1.1	0.0	1.1	100.0

Refer to question K_2 of the questionnaire for the original question

Table 5.50 - Number of Full-time Employees in 1992

It is interesting to note that 67.4% of the respondents had employed fewer than 5 (including zero) full-tim
employees at start-up, but by 1992, this percentage had reduced to 53.1%. Thus, the majority had
employed fewer than 5 (including zero) full-time employees at start-up, as well as in 1992. The reduction
the size of the majority over the life of the firms represents growth in numbers of full-time employees, w
increases occurring mainly in the '6 to 10' and '11 to 15' categories. The increases in these two categorie
were relatively evenly spread across the different types of contractors.

(b) Part-time employees: Tables 5.51 and 5.52 contain the data pertaining to part-time employees. As wa
the case for full-time employees, a considerable number of respondents reported having had no part-time
employees, both at start-up (38.9%) and in 1992 (37.5%). Over the life of the firms, there was a reductic
in the number of respondents who had employed only one skilled part-time employee - from 17.9% at sta
up to 8.3% in 1992. A slight increase occurred in the '2 to 5' category, but the '6 to 10' category had
doubled by 1992. No movement was recorded in the numbers of employees in the categories for greater
than 10 employees.

Count Row % Col % Total %	NONE	1	2 TO 5	6 TO 10	11 TO 15	16 TO 20	21 TO 30	31 TO 40	41 TO 50	MORE THAN 50	Row Total %
GENERAL	8	2	12	5	1	0	0	0	0	0	2
	28.6	7.1	42.9	17.9	3.6	0.0	0.0	0.0	0.0	0.0	
	21.6	11.8	38.7	55.6	100.0	0.0	0.0	0.0	0.0	0.0	
	8.4	2.1	12.6	5.3	1.1	0.0	0.0	0.0	0.0	0.0	
SPECIAL	4	4	6	0	0	0	0	0	0	0	1
	28.6	28.6	42.9	0.0	0.0	0.0	0.0	0.0	0.0	0.0	
	10.8	23.5	19.4	0.0	0.0	0.0	0.0	0.0	0.0	0.0	
	4.2	4.2	6.3	0.0	0.0	0.0	0.0	0.0	0.0	0.0	
LAB SUB	10	5	5	3	0	0	0	0	0	0	2
	43.5	21.7	21.7	13.0	0.0	0.0	0.0	0.0	0.0	0.0	
	27.0	29.4	16.1	33.3	0.0	0.0	0.0	0.0	0.0	0.0	
	10.5	5.3	5.3	3.2	0.0	0.0	0.0	0.0	0.0	0.0	
L&M SUB	15	6	8	1	0	0	0	0	0	0	3
	50.0	20.0	26.7	3.3	0.0	0.0	0.0	0.0	0.0	0.0	
	40.5	35.3	25.8	11.1	0.0	0.0	0.0	0.0	0.0	0.0	
	15.8	6.3	8.4	1.1	0.0	0.0	0.0	0.0	0.0	0.0	
Column Total %	37 38.9	17 17.9	31 32.6	9 9.5	1 1.1	0 0.0	0 0.0	0 0.0	0 0.0	0 0.0	1C

Refer to question K_1_2 of the questionnaire for the original question

Table 5.51 - Number of Part-time Employees at Start-up

It is of interest to note that 89.4% of the firms had employed fewer than 5 (including zero) part-time
employees at start-up, but by 1992, this percentage had reduced to 81.2%. However, at start-up, the
majority (56.8%) had employed zero or 1 part-time person. By 1992, this percentage had reduced to
45.8%. This indicates growth in numbers of part-time employees, with increases occurring mainly in the
to 10' category. The increase in this category was largely the result of increases reported by specialist an

labour-and-material sub-contractors.

Count Row % Col % Total %	NONE	1	2 TO 5	6 TO 10	11 TO 15	16 TO 20	21 TO 30	31 TO 40	41 TO 50	MORE THAN 50	Row Total %
GENERAL	13 44.8 36.1 13.5	1 3.4 12.5 1.0	9 31.0 26.5 9.4	5 17.2 29.4 5.2	1 3.4 100.0 1.0	0 0.0 0.0 0.0	0 0.0 0.0 0.0	0 0.0 0.0 0.0	0 0.0 0.0 0.0	0 0.0 0.0 0.0	29 30.2
SPECIAL	7 50.0 19.4 7.3	1 7.1 12.5 1.0	3 21.4 8.8 3.1	3 21.4 17.6 3.1	0 0.0 0.0 0.0	0 0.0 0.0 0.0	0 0.0 0.0 0.0	0 0.0 0.0 0.0	0 0.0 0.0 0.0	0 0.0 0.0 0.0	14 14.6
LAB SUB	10 43.5 27.8 10.4	4 17.4 50.0 4.2	4 17.4 11.8 4.2	5 21.7 29.4 5.2	0 0.0 0.0 0.0	0 0.0 0.0 0.0	0 0.0 0.0 0.0	0 0.0 0.0 0.0	0 0.0 0.0 0.0	0 0.0 0.0 0.0	23 24.0
L&M SUB	6 20.0 16.7 6.3	2 6.7 26.0 2.1	18 60.0 52.9 18.8	4 13.3 23.5 4.2	0 0.0 0.0 0.0	0 0.0 0.0 0.0	0 0.0 0.0 0.0	0 0.0 0.0 0.0	0 0.0 0.0 0.0	0 0.0 0.0 0.0	30 31.3
Column Total %	36 37.5	8 8.3	34 35.4	17 17.7	1 1.0	0 0.0	0 0.0	0 0.0	0 0.0	0 0.0	96 100.0

Refer to question K_2_2 of the questionnaire for the original question

Table 5.52 - Number of Part-time Employees in 1992

The following is a summary of the results regarding numbers of full-time and part-time employees :

(i)   the majority of respondents had employed fewer than 5 full-time and fewer than 5 part-time employees, both at start-up as well as in 1992.

(ii)   these majorities included significant numbers of firms which had never had employees.

(iii)   numbers of full-time and part-time employees grew over the life of the firms, but this growth was insufficient for the 1992 majority to represent a higher category of employee numbers.

It must be noted that these numbers of employees are extremely low. As noted in section 2.4.1, 'small' was defined as fewer than 20 employees. The data in this section indicate that the majority of the survey sample could better be described as 'micro', since it contained many firms with ten or fewer employees as well as many that were run solely by the owner. Although growth in numbers of employees over time was slight, the above findings tend to support Krafchik's (1990:127) observation that SSCEs have potential to absorb labour. In addition, the argument of section 2.3.3 (c) that SSCEs are potentially strong generators of employment is supported.

(c) Skilled employees: An analysis was performed (on the data in Tables 5.49 to 5.52) to establish what percentage of the full-time employees were qualified artisans - at start-up as well as in 1992. The results of

this analysis are presented in Tables 5.53 and 5.54, below.

Count Row % Col % Total %	0%	1% TO 10%	11% TO 20%	21% TO 30%	31% TO 40%	41% TO 50%	51% TO 60%	61% TO 70%	71% TO 80%	81% TO 100%	Row Total
GENERAL	12 48.0 29.3 17.6	1 4.0 100.0 1.5	4 16.0 50.0 5.9	2 8.0 50.0 2.9	2 8.0 33.3 2.9	1 4.0 25.0 1.5	0 0.0 0.0 0.0	0 0.0 0.0 0.0	2 8.0 100.0 2.9	1 4.0 50.0 1.5	
SPECIAL	9 81.8 22.0 13.2	0 0.0 0.0 0.0	1 9.1 12.5 1.5	0 0.0 0.0 0.0	0 0.0 0.0 0.0	1 9.1 25.0 1.5	0 0.0 0.0 0.0	0 0.0 0.0 0.0	0 0.0 0.0 0.0	0 0.0 0.0 0.0	
LAB SUB	6 60.0 14.6 8.8	0 0.0 0.0 0.0	1 10.0 12.5 1.5	2 20.0 50.0 2.9	1 10.0 16.7 1.5	0 0.0 0.0 0.0	0 0.0 0.0 0.0	0 0.0 0.0 0.0	0 0.0 0.0 0.0	0 0.0 0.0 0.0	
L&M SUB	14 63.6 34.1 20.6	0 0.0 0.0 0.0	2 9.1 25.0 2.9	0 0.0 0.0 0.0	3 13.6 50.0 4.4	2 9.1 50.0 2.9	0 0.0 0.0 0.0	0 0.0 0.0 0.0	0 0.0 0.0 0.0	1 4.5 50.0 1.5	
Column Total %	41 60.3	1 1.5	8 11.8	4 5.9	6 8.8	4 5.9	0 0.0	0 0.0	2 2.9	2 2.9	

Refer to question K_1 of the questionnaire for the original question

Table 5.53 - Skilled Employees as Percentage of Total Full-time Employees at Start-up

Count Row % Col % Total %	0%	1% TO 10%	11% TO 20%	21% TO 30%	31% TO 40%	41% TO 50%	51% TO 60%	61% TO 70%	71% TO 80%	80% TO 100%	Row Total
GENERAL	12 48.0 30.8 16.9	2 8.0 28.6 2.8	4 16.0 44.4 5.6	5 20.0 55.6 7.0	0 0.0 0.0 0.0	1 4.0 50.0 1.4	1 4.0 100.0 1.4	0 0.0 0.0 0.0	0 0.0 0.0 0.0	0 0.0 0.0 0.0	
SPECIAL	7 53.8 17.9 9.9	3 23.1 42.9 4.2	1 7.7 11.1 1.4	0 0.0 0.0 0.0	2 15.4 66.7 2.8	0 0.0 0.0 0.0	0 0.0 0.0 0.0	0 0.0 0.0 0.0	0 0.0 0.0 0.0	0 0.0 0.0 0.0	
LAB SUB	8 72.7 20.5 11.3	0 0.0 0.0 0.0	0 0.0 0.0 0.0	2 18.2 22.2 2.8	1 9.1 33.3 1.4	0 0.0 0.0 0.0	0 0.0 0.0 0.0	0 0.0 0.0 0.0	0 0.0 0.0 0.0	0 0.0 0.0 0.0	
L&M SUB	12 54.5 30.8 16.9	2 9.1 28.6 2.8	4 18.2 44.4 5.6	2 9.1 22.2 2.8	0 0.0 0.0 0.0	1 4.5 50.0 1.4	0 0.0 0.0 0.0	1 4.5 100.0 1.4	0 0.0 0.0 0.0	0 0.0 0.0 0.0	
Column Total %	39 54.9	7 9.9	9 12.7	9 12.7	3 4.2	2 2.8	1 1.4	1 1.4	0 0.0	0 0.0	

Refer to question K_2 of the questionnaire for the original question

Table 5.54 - Skilled Employees as Percentage of Total Full-time Employees in 1992

In both Tables 5.53 and 5.54 it can be seen that the majority of firms had no qualified artisans on their staff, both at start-up as well as in 1992. This is an important result. It suggests that skilled employees are employed on a part-time or casual basis, or that, the firm's owner is its only qualified artisan. Alternatively, this result might indicate that firms use unskilled personnel to perform skilled work. Since this was found by both Krafchik (1990:126) and Merrifield (1992:68), these data are taken as confirming those findings.

Notwithstanding the above, it must be noted that growth occurred in the percentages of qualified artisans, between start-up and 1992. This is indicated by the reduction in the size of the majority that employed no qualified artisans. This growth generally involved increases in the 1% to 30% categories - this can be seen from the fact that 14 firms recorded percentages between 31% and 100% at start-up, but only 7 firms recorded percentages in this range in 1992.

(d) Casual employees: The majority (59.4) of firms sometimes use casual labour. A relatively low 18.8% of the respondents indicated that they always use casuals, while 21.9% reported that the never use casuals (pg.308/AppH;Q-K_3). A majority (two-thirds) of the 75 respondents who answered the question on average number of casuals employed at a time, reported having employed fewer than 4 (pg.309/AppH;Q-K_3_1).

5.11.2 Employee Remuneration

In the following two tables (Tables 5.55 and 5.56), details are given regarding the wages or salaries paid to employees. The data apply to all types of contractors, but Appendixes K1 and K2 may be consulted for more detail on wages and salaries paid by the various types of contractors.

In considering Table 5.55, it must be noted that the monthly wages for 'labourer' in Table 5.55 appear to be incorrect - 11 of the 70 respondents noted that they paid labourers more than R1 200 per month. Although possible, this is very unlikely. The data is therefore potentially unreliable and for this reason analysis has been kept to a minimum.

Notwithstanding the possibility that the data are unreliable, it is of interest to compare the survey results with the average wage paid to black construction workers. Table C6 in Appendix C shows that the 1989 average wage for this group was R592 per month. This converts to a constant 1992 value of R900 (using the CPI). This figure of R900 can then be compared with the constant 1992 values for the survey data contained in Table 5.56. Before drawing any conclusions from this comparison, it must be noted that the figure of R900 represents a weighted average for all types of employee. The survey data have not been averaged, since this would require specific data on the average number of employees of all types employed during 1991. This data was not requested in the survey.

Count Row %	LESS THAN R300	R301 TO R600	R601 TO R900	R901 TO R1 200	R1 201 TO R1 500	R1 501 TO R1 800	R1 801 TO R2 100	R2 101 TO R2 400	R2 401 TO R2 700	MORE THAN R2 700	Row Total %
					PER MONTH						
CLEANER	4 14.8	16 59.3	5 18.5	1 3.7	1 3.7	0 0.0	0 0.0	0 0.0	0 0.0	0 0.0	
LABOURER	5 7.1	25 35.7	21 30.0	8 11.4	3 4.3	2 2.9	2 2.9	1 1.4	2 2.9	1 1.4	
APPRENTICE	1 6.3	4 25.0	6 37.5	2 12.5	1 6.3	1 6.3	1 6.3	0 0.0	0 0.0	0 0.0	
QUALIFIED ARTISAN	0 0.0	4 12.1	6 18.2	9 27.3	4 12.1	0 0.0	2 6.1	2 6.1	3 9.1	3 9.1	
FOREMAN	1 3.8	3 11.5	4 15.4	7 26.9	4 15.4	3 11.5	1 3.8	0 0.0	1 3.8	2 7.7	
CLERK	5 15.6	5 15.6	15 46.9	4 12.5	2 6.3	0 0.0	0 0.0	1 3.1	0 0.0	0 0.0	

Refer to question K_5 of the questionnaire for the original question

Table 5.55 - Wages and Salaries Paid to Employees in 1991

TYPE OF EMPLOYEE	AVERAGE MONTHLY WAGE/SALARY	
	SURVEY RESULTS (1991 VALUES)	SURVEY RESULTS (CONSTANT 1992 VALUE...)
CLEANERS	R528.01	R60...
LABOURERS	R834.87	R96...
APPRENTICES	R859.63	R98...
QUALIFIED ARTISANS	R1 499.14	R1 72...
FOREMEN	R1 292.02	R1 48...
CLERKS	R754.06	R86...

N.B.   Constant 1992 values were derived by applying the Consumer Price Index to the survey data.

Table 5.56 - Average Monthly Wages/Salaries Paid to Employees in 1992

From a study of Table 5.56, it is clear that there are no significant differences between the monthly wag...
for labourers (R960), apprentices (R988) and clerks (R867) when these are compared with the average f...
all construction industry Blacks (R900). However, the wages for cleaners (R607) are lower, while for
artisans (R1 724) and foremen (R1 485) they are higher. These wages which, respectively, are lower an...
higher would, of course, affect the weighted average wage. For example, a team consisting of 1 artisan,
1 apprentice and 4 labourers would have a weighted average wage of R1 092 per month, using the 1992
values from the survey data. This example is speculative and cannot be applied to the survey data with a...
confidence. What can be noted, however, is that in all categories of employee, except 'cleaners' and

'clerks', the monthly wages exceed the R900 industry average. This suggests that the employees of the survey respondents are above average earners. This is supported, in the case of the 'labourer' category, by the comparison of their wages (R960) with the Industrial Council minimum wages (see Appendix B(c)) which range between R608 and R659. However, in the case of the 'artisan' category, the Industrial Council minimum wages range between R1 818 and R1 941 (see Appendix B(c)) and these amounts both exceed the R1 724 recorded in the survey for this group.

As already noted, these data could be unreliable. This tends to be confirmed by Krafchik's (1990:133-137) findings that the employees of his sample of sub-contractors were paid substantially less than the statutory minimum wages. A possible explanation for the apparently incorrect amounts presented in Table 5.56 is that the labour category names were interpreted loosely by the respondents.

5.11.3 Extent of Sub-contracting

The following table (Table 5.57) contains the responses to the questions put to general contractors only, regarding whether or not they employ sub-contractors to do various elements of their work. Although the survey included 29 general contractors, only 26 responded to the questions on labour-only sub-contracting, and only 25 to those on labour-and-material sub-contracting.

The results presented in Table 5.57 are important for three reasons. Firstly, the majority of respondents never employ sub-contractors - neither on a labour-only, nor on a labour-and-material basis. Secondly, where sub-contracting occurs, labour-only sub-contracting is far more common than labour-and-material sub-contracting. Thirdly, the majority of respondents reported that they 'never' sub-contracted plumbing and electrical work to others. Regarding the latter point, this either means that most general contractors have qualified plumbers and electricians in their employ, or that they do this work themselves (probably without the necessary qualifications). Another possibility is that they employ plumbers and electricians as individuals, on a part-time or casual basis. Thus, in answering the question, they could have considered this practice to be that of *employing*, rather than *sub-contracting*.

These data further indicate that black sub-contractors do not readily obtain employment through black general contractors. And it has been noted that black SSCEs are generally locked into black residential markets (see sections 2.5.2 and 5.7.2). If this is accepted, it must mean that the constraint of lack of access to work opportunities is more severe for black sub-contractors than it is for black general contractors.

Count Row %	EMPLOY LABOUR-ONLY SUB-CONTRACTOR				EMPLOY LABOUR & MATERIAL SUBCONTRACTOR			
GENERAL CONTRACTORS ONLY	Always	Sometimes	Never	Row Total %	Always	Sometimes	Never	Row Total %
BRICKWORK	4 15.4	2 7.7	20 76.9	26 100.0	2 8.0	1 4.0	22 88.0	100.0
ROOF CARPENTRY	4 15.4	3 11.5	19 73.1	26 100.0	1 4.0	4 16.0	20 80.0	100.0
ROOF TILING	3 11.5	4 15.4	19 73.1	26 100.0	1 4.0	1 4.0	23 92.0	100.0
PLASTERING & SCREEDING	5 19.2	4 15.4	17 65.4	26 100.0	3 12.0	4 16.0	18 72.0	100.0
PAINTING	6 23.1	7 26.9	13 50.0	26 100.0	3 12.0	5 20.0	17 68.0	100.0
CEILINGS	4 15.4	1 3.8	21 80.8	26 100.0	1 4.0	3 12.0	21 84.0	100.0
GLAZING	5 19.2	1 3.8	20 76.9	26 100.0	4 16.0	1 4.0	20 80.0	100.0
FLOOR COVERINGS	5 19.2	3 11.5	18 69.2	26 100.0	2 8.0	5 20.0	18 72.0	100.0
PLUMBING	1 3.8	6 23.1	19 73.1	26 100.0	2 8.0	1 4.0	22 88.0	100.0
ELECTRICAL	6 23.1	4 15.4	16 61.5	26 100.0	4 16.0	5 20.0	16 64.0	100.0

Refer to question K_4 of the questionnaire for the original question

Table 5.57 - Extent to which General Contractors Employ Sub-contractors

5.11.4 Registration with Industrial Council

Readers not well acquainted with the Industrial Council system should consult Appendix B for details. The responses to the question on whether or not the firm was registered with the Industrial Council reveal an important finding, that is, that only 38.5% of the firms were registered (pg.342/AppH;Q-K_6). This means that the majority (61.5%) of firms were effectively operating illegally. This suggests that the majority of firms are essentially 'informal' enterprises (as defined in section 3.4.4). A further finding was that 38.5% of the respondents reported that it was impossible for Industrial Council inspectors to check on them, and a further 14.6% reported that this would be difficult (pg.344/AppH;Q-K_7). The data thus support previous findings that the majority of black SSCEs fall within the informal sector.

The following table (Table 5.58) contains the responses of those who gave reasons regarding why they have not registered with the Industrial Council.

Count	Col %	NUMBER	Column %
I don't take much interest in laws and regulations because the authorities make trouble for Blacks	1	1.7
Don't know where to go or how to get registered	4	6.9
Just started firm - still informal - will register when I get more professional/bigger	9	15.5
Too expensive	6	10.3
Blacks don't have to register if working in black areas - only whites working in white areas	1	1.7
If registered I would have to pay more government taxes - I want to avoid this	6	10.3
I thought I needed a qualification before I could register	1	1.7
There would be too many regulations to comply with if I registered	1	1.7
Have not yet thought of registering	4	6.9
Am in the process of registering	3	5.2
Blacks are not allowed to register - only Whites	1	1.7
Never heard of the Industrial Council / don't know what it is	7	12.1
Do not intend to register	1	1.7
Can't see the benefit	3	5.2
Do not own property / not a South African citizen	2	3.4
Because I am not registered as a company / business	2	3.4
Because I don't have a license	1	1.7
Because I am too inexperienced - will register when I have received more training	1	1.7
No response	4	6.9
Column	Total %	58	100.0

Refer to question K_6_1 of the questionnaire for the original question

Table 5.58 - Reasons for Not Registering with Industrial Council

Most of the responses are self-explanatory and should simply be read from the table. The most common response (from 15.5% of the sample of 58) was that the respondent though his firm was too small to be registered, or that it was not yet professional enough. Alarmingly, the second most common response (from 12.1% of the 58) was that the respondent was unaware that the Industrial Council exists.

Two other responses, totalling 20.6% (each came from 10.3% of the sample of 58) reveal the most common problem - expense. Notably, half of these respondents referred to 'government taxes', rather than 'expense'. Thus, the perception exists that Industrial Councils are channels for government revenue, rather than the administrators of employer-employee agreements. It should also be noted that 3 (5.2% of the 58) individuals made reference to issues involving racial discrimination. It is likely that other respondents also harboured such impressions, but did not report them.

The findings of this section essentially show the Industrial Council system to be unworkable. Its existence in its current form leaves the majority of *bona fide* small contractors in default of the law and it effectively gives rise to the informal sector. Perhaps the system's most inappropriate attribute is that it is also binding on people other than the members of the employer and employee bodies that formed the various agreements (see section 2.5.2 (a)).

This section contains the analysis of the problems experienced by the respondents in operating their firms
The data presented in this section concern only what were described by the respondents as *serious*
problems. The method employed to filter out these serious problems was as follows. Module L of the
questionnaire required respondents to rank, on a scale of 1 (low) to 5 (high), a list of 23 problems falling
under the categories: Labour; Finance; Production/Management; Legislation/Government; and Markets.
The results of this ranking are presented without any analysis on pages 345 to 367 of Appendix H
(questions L_1_1 to L_1_23). The reason for these data not being analysed is that questions L_2 to L_6 o
the questionnaire required respondents to identify the five most serious problems from this batch of data
and it is these more serious problems that are analysed in section 5.12.2.

5.12.1 Serious Problems

To identify serious problems, respondents were then asked to select five of the problems that they had
ranked '4' or '5'. They were then asked to explain the exact nature of the problem, as well as what woul
improve it. For the purposes of the following analysis, these problems are referred to as problems 'A' to
'E'. Problem A was the first mentioned of the five given by each respondent. Problem B was the second
mentioned, and so on. A perusal of the original questionnaires revealed that problems A to E were usuall
the five highest ranked problems, presented *in the order of the list in the questionnaire*. It is thus importa
to note that, of the five problems, problem A was not necessarily the most important, and problem E the
least important. Table 5.59 summarises the number of responses obtained for each of the problems A to I
cross tabulated by problem label.

The second column (from the left) of Table 5.59 lists the problems given in the questionnaire. A total for
each of these problems is given in the right-hand column of the table. This total represents the number of
respondents who cited that particular problem as one of their five most serious constraints. Thus, as has
been done in Table 5.59, the problems can be ordered or ranked according to this total. The column total
for each problem (A to E) represents the number of the 96 respondents who answered questions L_2 to
L_6. Thus, 80.2% of the respondents reported having a problem A (that is, answered question L_2), 79.
reported having a problem B (that is, answered question L_3), *etc.*. The most common reason for
questions L_2 to L_6 not having been answered was that the interviewee refused to complete the intervie
(too time-consuming), but in addition, some respondents had not ranked enough problems '4' or '5' to be
able to name five serious problems.

Count		FIVE PROBLEMS WHICH WERE RANKED '4' OR '5'					
ORDER	SERIOUS PROBLEMS (in order of frequency)	A	B	C	D	E	Total
1	Access to loans/security	11	22	6	0	0	39
2	Cash flow/slow payment	1	4	9	15	1	30
3	Supply/cost of materials	2	9	10	4	0	25
4	Cost of labour	23	0	0	0	1	24
5	Shortage of skilled labour	17	4	2	0	0	23
5	Preference for white contractors	1	2	3	9	8	23
6	Lack of technical skills	3	4	9	1	1	18
7	Reliability of labour	9	7	0	0	1	17
8	Competition	0	0	5	2	9	16
9	Access to land	2	1	4	2	6	15
10	Relationship with main contractor	3	6	1	0	0	10
11	Cost of/access to equipment	1	2	5	1	0	9
12	Worker action	1	6	1	0	0	8
12	Tax legislation	1	0	2	1	4	8
13	Working out total costs	1	1	2	3	0	7
14	Tender/negotiation procedures	0	2	2	1	1	6
15	Accounting/reconciliation	1	1	2	0	1	5
16	Repayment of loans	0	1	2	1	0	4
16	Erratic inflow of work	0	1	2	0	1	4
17	Industrial Councils/labour legislation	0	0	2	0	1	3
18	Relationship with sub-contractors	0	2	0	0	0	2
18	Interest rate issues	0	1	1	0	0	2
N/A	Other legislation	0	0	0	0	0	0
	Column total	77	76	70	40	35	298
	% of survey total	80.2	79.2	72.9	41.7	37.5	62.1*

Refer to questions L_1_1 to L_1_23 of the questionnaire for the original questions

* This percentage is calculated as follows: 298÷480, where 480 = (5 x 96) which was the total possible number of responses.

N.B. This table must be read in conjunction with Table 5.82 which contains a revision of its contents.

Table 5.59 - Serious Problems

As will be shown in section 5.12.2, the analysis of the nature of, and solutions to, the serious problems revealed that some of the problem labels had clearly been misinterpreted by certain interviewees. Table 5.59 has thus been amended in the light of this analysis and has been re-compiled as Table 5.82.

5.12.2 Nature of and Solutions to Serious Problems

The data contained in the tables in this section flow directly from Table 5.59. For example, 39 respondents indicated 'access to loans/security' as a serious problem (see 'total' column). However, they were also asked to explain the problem and suggest what would improve it. The responses to these questions are presented in Table 5.60. Similarly, the data pertaining to the other problems listed in Table 5.59 are presented in the tables numbered 5.61 to 5.71 in this section, and 5.72 to 5.81 in Appendix Q. (Tables 5 to 5.81 involve small sample sizes and relatively few responses regarding the nature of and solutions to problems. They are therefore presented in Appendix Q (without any analysis of their contents).

The descriptions in Tables 5.61 to 5.81 regarding the nature of and solutions to the problems were compiled by the author and are a summary of the actual responses. Appendixes M2 and M1, respectively should be consulted for the actual responses and the summary of these into the labels used in Tables 5.6 5.81.

(a) The 'access to loans/security' problem: The <u>inability to raise collateral</u> was the most common reason why 'access to loans/security' was cited as a problem. This applied to 48.7% of the 39 respondents.

ACCESS TO LOANS/SECURITY PROBLEM		
NATURE OF PROBLEM	COUNT	
Loans not granted because I am too old	1	
Financial institutions will not lend to Blacks	6	
Difficult to qualify for loan because can't raise security	19	
Loans are not granted to unregistered businesses	2	
Do not know how to apply	2	
Only well-know and successful people get loans	4	
Balance sheets are required with the application - I can't afford to have them prepared	1	
No response or don't know	4	
SOLUTION TO PROBLEM		COUNT
Black contractors' associations should present the needs of their members to the financial institutions		4
Banks should relax rules for loan applications from Blacks		15
Apartheid must stop - Blacks must be allowed to qualify for loans		4
Security should not be required until the business is established		3
Become better educated/ experienced - then loans will easily be obtained		2
Nothing		1
I must apply when I know how to		1
No response or don't know		9
TOTAL	39	39

The findings in Table 5.60 are related to, and should be read in conjunction with those presented in section 5.10.2.

Table 5.60 - Nature of and Solutions to ACCESS TO LOANS/SECURITY Problem

Six (6) (15.4% of the 39) respondents believed that only registered or well-known businesses qualified for loans. A further 6 respondents (15.4% of the 39) believed that they had been refused loans because they were black. This is an important result because it suggests that the respondents believed that a white applicant, with exactly the same track record and ability to raise collateral, would have successfully secured a loan. This is unlikely to be true.

In response to the question regarding what would improve the problem, the <u>majority of respondents felt that loan application criteria should be less stringent for black applicants</u>. This implies a lack of understanding of why financial institutions impose the loan application criteria that they do, and that all applicants, regardless of race, would be subjected to the same criteria.

(b) The 'cash flow/slow payment' problem: Table 5.61 reveals that <u>cash flow problems are precipitated by clients</u> who pay later than the originally agreed time, or less than the agreed amount. This was reported by 21 (70.0% of the 30) respondents.

CASH FLOW/SLOW PAYMENT PROBLEM		
NATURE OF PROBLEM	COUNT	
Clients pay too late and often less than agreed	21	
My costs increase due to late payment	4	
Slow payment makes it hard to calculate profits and plan business	4	
No response or don't know	0	
SOLUTION TO PROBLEM		COUNT
Clients should have loan facilities to be able to pay 60% to 80% in advance		3
Slow payment should not be allowed - people should be honest		5
Joining an association which could help follow up slow payers		1
Getting signed stop orders from clients		2
Department of manpower must solve the problem		1
An improvement in the economy		2
Should make clients obtain loans from banks and should not accept cash from them		2
Clients should pay whole contract amount in advance		4
Clients should be subsidised		1
Lawyers should be involved in signing agreements		3
We should only work for cash		1
People should not be allowed to occupy their homes before they have paid		1
No response or don't know		4
TOTAL	30	30

Table 5.61 - Nature of and Solutions to CASH FLOW/SLOW PAYMENT Problem

The proposed solutions to this problem were numerous and varied. A relatively common suggestion was that <u>clients should pay in advance, either all of, or a high percentage of the contract amount</u>. This is an uncommon practice in the 'formal' building industry, where a contractor would usually only receive the first payment after one month, and thereafter would receive monthly payments equal to the value of work *satisfactorily* completed.

A similar observation to that made about the proposed solutions to the 'access to loans/security' problem, can be made of these responses - the solutions tend to be unrealistic and propose that rules be relaxed to accommodate the capabilities of the respondents. Certain of the proposed solutions point to the problem of the inability of clients to secure mortgage loan finance or housing subsidies. This is an important revelation since it indicates that, to a degree, the problems of the respondents are beyond their own control. Indeed, noted in (section 2.5.2 (d) and Chapter 4) financial institutions have recently tended to shy away from the black home loan market because of the high risk of bond boycotts and repayment problems.

(c) The 'supply/cost of materials' problem: Notably, none of the responses to the question regarding the nature of the 'supply/cost of materials' problem indicated that it involved the *supply* of materials to sites.

SUPPLY/COST OF MATERIALS PROBLEM		
NATURE OF PROBLEM	COUNT	
Materials are expensive - lack capital to finance their purchase	3	
We black contractors do not get discounts	10	
Too expensive, mainly due to tax	12	
No response or don't know	0	
SOLUTION TO PROBLEM		COUNT
Clients should pay in advance		2
Discounts should be given to all contractors		9
We should be allowed to buy in bulk		1
Black builders' associations should establish buying clubs or negotiate with suppliers		3
Nothing can improve the situation		1
Suppliers should give us credit - to be paid back in instalments		1
Material prices should be lowered		3
VAT should be lowered or done away with		2
Income tax should be lowered		1
The situation would improve if I had access to loans		1
Government should improve its running of the economy		1
No response or don't know		0
TOTAL	25	25

Table 5.62 - Nature of and Solutions to SUPPLY/COST OF MATERIALS Problem

Most commonly, the responses in Table 5.62 show the problem to be that <u>materials are too expensive</u> and that <u>black contractors do not get discounts</u>. Together, these responses accounted for 22 of the 25 responses (88.0%).

A contractor would usually recover the full cost of materials from his client, regardless of their cost. If this were the case, the problems of lack of access to discounts and expensive materials would essentially be problems of *lack of bridging finance*. However, since competition among SSCEs was reported to be fierce (see Table 5.68), it is likely that many would agree to build for less profit, or even at a loss. Indeed, the findings in Table 5.24 suggest that this is true - the majority negotiate their contracts, which probably means that they lower their original amounts in order to secure the job. Thus, it is more likely that the problem of expensive materials does, indeed, refer to the high cost of materials, than to the shortage of bridging finance. It must be noted that 10 of the 25 respondents (40.0%) believed that the reason they failed to qualify for discounts was because they are Black. <u>Three (3) responses identified lack of bridging finance to be the real problem</u>.

The most commonly proposed solution to the problem (by 9 of the 36, or 36.0%) was that <u>all contractors (that is, both Black and White) should enjoy discounts from materials suppliers</u>. However, 4 of the 25 responses (16.0%) suggested that access to bulk buying facilities would improve the problem. Most of these indicated that the CAs could play a role in facilitating this, which tends to support the argument in Chapter 4 that CAs fail to provide what members need.

(d) The 'cost of labour' problem: 83.3% of the 24 respondents reported that the problem was that <u>labour was too expensive</u> (see Table 5.63, below). As was noted about materials in the comments to Table 5.62, the full cost of labour would usually be recovered from the client. Why then would the cost of labour appear excessive to the contractor? A tentative answer is that this finding might indicate that *lack of bridging finance* is the real problem, but this cannot be established with certainty. It is possible that the competitive tendering market forces contractors to underquote on labour costs. But, it must be noted that <u>labour costs</u> would appear excessively high in cases where the contractor was unable to estimate productivity, and consequently experienced labour costs in excess of what had been expected. None of the respondents reported this to be the case, but that alone does not discount it as a possibility. While these points have explored reasons why labour costs might appear to be more expensive than they actually are, it must be remembered that in section 5.11.2, the survey data also revealed labour costs that appeared excessively high.

Two types of responses emerged in the section of the table dealing with solutions to the problem. The first type of response (accounting for 37.5% of the 24 responses) generally indicated that <u>access to bridging finance would improve the problem</u>. It is also evident from the response of 'get more jobs / increase income / charge clients more', that deposits and income from contracts are a source of bridging finance (this confirms the suggestion made in section 2.5.2 (d) that deposits are a substitute for formal loans). In addition to these responses, 3 (12.5% of the 24) respondents suggested that they could improve the problem by <u>cutting overhead costs and employing casual, instead of full-time, workers</u>. This tendency, however, would have a detrimental impact on the ability of the firms to establish dependable and efficient work forces.

COST OF LABOUR PROBLEM		
NATURE OF PROBLEM	COUNT	
Labour too expensive	20	
Labourers later demand more than was originally agreed upon	1	
Forced to underpay labourers because employer pays me too little	1	
Labour demand more than skills are worth	2	
No response or don't know	0	
SOLUTION TO PROBLEM		COUNT
Reduce size of labour force and employ more casuals		3
Wages should be agreed and fixed before job starts		1
Get more jobs / increase income / charge clients more		4
Disband labour unions		1
Create more money to pay labour by suppliers decreasing costs of materials and equipment		1
Employ properly skilled labour		2
Labour must agree to allow wages to fluctuate		2
Having access to loans		2
Education/training should be compulsory		1
I should be subsidised so that I can pay labourers properly		2
No response or don't know		5
TOTAL	24	24

Table 5.63 - Nature of and Solutions to COST OF LABOUR Problem

The second type of response (accounting for 16.6% of the 24 responses) suggested that <u>training for work would improve the problem</u>, or that <u>only properly skilled workers should be employed</u>. The latter suggestion must be viewed in the light of the data in the following table (Table 5.64) where it will be shown that skilled labour is in short supply.

(e) The 'shortage of skilled labour' problem: From Table 5.64 it can clearly be seen that there were two main reasons why 'shortage of skilled labour' was cited as a problem. Firstly, 9 (39.1% of the 23) respondents noted that there was an <u>undersupply of skilled workers</u>. This response was expected since it was shown in Tables 5.52 and 5.53 that the majority of firms had no skilled employees on their full-time staff. A further 2 (8.7% of the 23) respondents reported that some 'skilled' workers are, indeed,

unqualified in those skills.

SHORTAGE OF SKILLED LABOUR PROBLEM		
NATURE OF PROBLEM	COUNT	
Labour is not sufficiently skilled to work without supervision	8	
Black labour is less skilled than white	1	
Skilled labourers claim to have skills that they don't have	2	
There is an undersupply of skilled labour	9	
Skilled labourers do not want to work for black builders	1	
No response or don't know	2	
SOLUTION TO PROBLEM		COUNT
Supervision on site		1
Bursaries should be available for matriculants wanting to enter the building industry as skilled workers		4
Labour must be trained at training centres		2
Labour must be trained at Technical Colleges		2
Training		2
Government must open free training schools		6
More training institutions should be opened offering 1 to 2 year courses		2
No response or don't know		4
TOTAL	23	23

Table 5.64 - Nature of and Solutions to SHORTAGE OF SKILLED LABOUR Problem

Secondly, 8 (34.5 of the 23) respondents explained that the problem was the <u>inability of their employees to work without supervision</u>. By implication, they believed that properly skilled workers would not require supervision.

There was consensus on what would improve these problems. This can be seen from Table 5.64, where 18 (78.3% of the 23) respondents indicated that <u>training in trade skills should be available to their employees</u>. The most common suggestion was that this <u>training should be provided free of charge</u> by the government.

(f) The 'preference for white contractors' problem: From Table 5.65, above, it can be seen that 11 (47.8% of the 23) respondents believed themselves to be <u>victims of the perception that Whites are better builders than Blacks</u>. A further 6 (26.1% of the 23) respondents reported that the <u>Township Councils were responsible for the monopolisation of work in the townships by white contractors</u>.

A variety of solutions to this problem were suggested. <u>Most of these suggested actions that should be taken by clients, Township Councils, government, white contractors, or contractors' associations</u>. Notably, only one respondent felt that an improvement in the quality of his work would improve matters.

PREFERENCE FOR WHITE CONTRACTORS PROBLEM		
NATURE OF PROBLEM	COUNT	
People think that whites build better than Blacks	5	
Whites monopolise work in the townships - Township Councils are to blame for this	7	
Black contractors cannot get work in the cities and white areas because people prefer big white companies	6	
Cannot compete with Whites	1	
People think that Whites are honest and that Blacks are dishonest	1	
No response or don't know	3	
SOLUTION TO PROBLEM		COUNT
People should support local township builders		2
Negotiations between black builders and Township Councils		1
Government must give us a chance to prove ourselves		1
Advertising my firm in the media		1
White developers should at least use black sub-contractors		1
Nothing		1
Contractors should be selected on merit and Blacks should be given a chance to prove themselves		5
Black builders' associations should buy sites for their members		1
Free open market		2
Black contractors must be more faithful to their clients		1
We need a multi-racial builders association		1
By improving the quality of my work		1
No response or don't know		5
TOTAL	23	23

Table 5.65 - Nature of and Solutions to PREFERENCE FOR WHITE CONTRACTORS Problem

An important aspect to certain of these responses is that the respondents believe black townships to be the rightful domain of black, rather than white, contractors. This probably stems from the fact that black contractors are almost entirely excluded from working in the white areas and that the townships are their only real market.

(g) The 'lack of technical skills' problem: The results presented in Table 5.66, below, are related to the findings of section 5.4.4 and should be read in conjunction therewith.

A majority of 10 (55.6% of the 18) respondents explained that they themselves lacked technical skills, while the remainder of 8 (44.4% of the 18) referred to the lack of technical skills among their employees. It must be noted that these 8 responses are similar to the responses in Table 5.64.

LACK OF TECHNICAL SKILLS PROBLEM		
NATURE OF PROBLEM	COUNT	
Employees lack technical skills	8	
Owner lacks technical skills	10	
No response or don't know	0	
SOLUTION TO PROBLEM		COUNT
More training courses and centres should be made available - especially for semi-literate people and must be after hours		6
Workers should get training		2
Workers should attend Technikon		3
I (firm owner) should go for further training		4
Black builders' associations should get the private sector to promote technical support for Blacks		2
There should be entrepreneurial training courses		1
No response or don't know		0
TOTAL	18	18

Table 5.66 - Nature of and Solutions to LACK OF TECHNICAL SKILLS Problem

Predictably, the proposed solutions to these problems all suggested that <u>further training</u> would improve them. Notably, 6 (33.3% of the 18) suggestions pointed to the need for <u>after-hours training or training for semi-literate individuals</u>.

(h) The 'reliability of labour' problem: In Table 5.67, below, it is seen that 10 (58.8% of the 17) responses identify the <u>inability of labour to work without supervision</u> as the problem. It must be noted that this was also cited as a major reason why 'shortage of skilled labour' was perceived to be a problem in Table 5.64.

The most common suggestion regarding what would improve the problem, was <u>supervision (by the owner or a foreman)</u>. In this regard, it must be noted that only 26 respondents provided data on foremen's salaries in Table 5.55. It is thus assumed that only 26 (27.1%) of the sample actually employed a foreman. Further, it was found in Table 5.56 that the average 1992 salary for a foreman was R1 486 per month. It is therefore likely that the <u>real problem is the inability to carry the overhead cost of a foreman</u>.

RELIABILITY OF LABOUR PROBLEM		
NATURE OF PROBLEM	COUNT	
Labourers assist gangs to steal materials / dishonesty	3	
Labour cannot work without supervision	10	
Working hours are too long - labourers get tired and quality of work deteriorates	1	
Workers often don't arrive for work	2	
No response or don't know	1	
SOLUTION TO PROBLEM		COUNT
Should have security guards on site		2
Need a foreman / supervision		6
Should only work 5 hours a day - workers should be trained during the other 3 hours		1
Labourers should be severely punished		1
Labour legislation should be abolished so we can treat them how we like		2
Employer and employee should be bound by a written contract		1
Labourers should be more honest		1
No response or don't know		3
TOTAL	17	17

Table 5.67 - Nature of and Solutions to RELIABILITY OF LABOUR Problem

(i) The 'competition' problem: It was found from Table 5.68, below, that the problem of 'competition' largely represented competition among black SSCEs, rather than between this group and large white contractors. Eleven (11), or 68.8% of the 16 respondents, found it problematic that <u>too many SSCEs were competing for the same market</u>.

A variety of solutions to the problem were suggested. Apart from the 4 (25.0% of the 16) responses that nothing could improve the problem, no pattern emerged among the suggestions. It should be noted that one respondent made reference to the <u>role played by the Township Councils in controlling the flow of work in the black areas</u>, implying that these bodies actually cause the problem.

COMPETITION PROBLEM		
NATURE OF PROBLEM	COUNT	
Many small contractors are competing for the same market	11	
Can't compete with the white developers	3	
No response or don't know	2	
SOLUTION TO PROBLEM		COUNT
Client should choose the contractor		1
Developer should sub-contract to local black contractors and not use outsiders		1
Move into larger area/market		1
Nothing		4
Township Councils should give Blacks a chance		1
We should be more negotiable on cost		1
Confidence in work and quality of work		1
Well trained salesmen sponsored by government		1
No response or don't know		5
TOTAL	16	16

Table 5.68 - Nature of and Solutions to COMPETITION Problem

(j) The 'access to land' problem: All of the 15 respondents indicated that, for one reason or another (see Table 5.69 below), that they <u>could not obtain sites</u>. This either suggests that they would be building for speculative purposes if they could obtain sites, or that their clients rely on them to locate sites. It is important to note that the problem of '... Apartheid and reservation of work for local contractors' does not necessarily refer to the, now defunct, Group Areas Act. Indeed, it appears that the term 'Apartheid' was used to refer to the preference of Whites to employ white contractors, while the term 'reservation of work' appears to refer to the actions of the Township Councils.

One-third of the suggested solutions involved <u>government intervention</u>. A further one-third suggested that <u>Township Councils should allocate a quota of sites to black contractors</u>. It is clear from the responses in Tables 5.64, 5.67 and 5.68 that the Township Councils are powerful bodies in terms of their control over the development of townships. Although it is unlikely that any contractors represented in this survey would be capable of undertaking a full township development, they nevertheless consider themselves capable of building the houses involved in such developments and consequently resent the presence of white developers in these areas.

ACCESS TO LAND PROBLEM		
NATURE OF PROBLEM	COUNT	
Big white developers get all the land	7	
Can't get sites	3	
Cannot work in certain areas because of Apartheid and reservation of work for local contractors	5	
No response or don't know	0	
SOLUTION TO PROBLEM		COUNT
Government should make it possible to operate anywhere, regardless of colour		4
Government must provide us with work		1
Township Councils should allocate 60% of sites to black contractors		1
Township authorities must give land to Blacks - not only white developers		4
Big white contractors and the government must end Apartheid and give us a chance		3
Government should abolish reservation of work for local contractors		1
Builders' association should buy land and let its members develop it		1
No response or don't know		0
TOTAL	15	15

Table 5.69 - Nature of and Solutions to ACCESS TO LAND Problem

(k) The 'relationship with main contractor' problem: Table 5.70, below, contains the responses to the questions on the nature of and solutions to the problem of the 'relationship with main contractor'. The sample size was relatively small, involving only 10 respondents, who, given the nature of the problem, must all have been sub-contractors. Four (4) of these 10 reported that they had either been 'exploited', or underpaid by main contractors. No pattern emerged from the proposed solutions to the problem.

(l) The 'cost of/access to equipment' problem: The data in Table 5.71, below, indicate that the majority (88.9% of the 9) respondents, could not afford to buy equipment, but had to hire it. Only one (11.1% of the 9) indicated that hire charges were too expensive. The proposed solutions to the problem are, in some cases, unrealistic. The hiring of equipment is a very common building industry practice, particularly among SSCEs. Without further detail on precisely which items of equipment were deemed unaffordable, it is impossible to assess the severity of the problems regarding the acquisition of equipment.

RELATIONSHIP WITH MAIN CONTRACTOR PROBLEM		
NATURE OF PROBLEM	COUNT	
Main contractor pays too little	2	
I am not know to the main contractors	1	
Black sub-contractors are exploited	2	
Main contractor demands that we work under pressure	1	
No response or don't know	4	
SOLUTION TO PROBLEM		COUNT
Should be paid according to job done - not certificate held		1
Lawyers should be involved to prevent exploitation		1
Main contractors should not pressurise us - we should be allowed to work at our own pace		1
Should make myself known to main contractors		1
Main contractors should stick to agreements		1
They should pay us better		1
No response or don't know		4
TOTAL	10	10

Table 5.70 - Nature of and Solutions to **RELATIONSHIP WITH MAIN CONTRACTOR** Problem

COST OF/ACCESS TO EQUIPMENT PROBLEM		
NATURE OF PROBLEM	COUNT	
Can't afford to buy equipment - must hire it	8	
Hire charges are too high	1	
No response or don't know	0	
SOLUTION TO PROBLEM		COUNT
Access to hiring institutions		1
Making enough profit to purchase my own equipment		1
Equipment prices should be lowered		3
Nothing		1
Discount should be offered to members of builders' associations		2
Should be able to include the full cost of equipment when charging the client		1
No response or don't know		0
TOTAL	9	9

Table 5.71 - Nature of and Solutions to **COST OF/ACCESS TO EQUIPMENT** Problem

5.12.3 Basis for Amendment of Table 5.59

As noted in section 5.12.1, Table 5.59 represented the responses to questions where the interviewees we
asked to identify five serious problems. A list of problems was provided and it was from this list that th
short description of the problem was taken. For example, the respondent might have indicated that:
'competition'; 'tax legislation'; 'worker action'; 'access to land'; and 'cost of labour'; had been serious
problems. However, the exact nature of each of the problems might not have been adequately
communicated by the predefined problem labels that formed the range of alternatives. Indeed, this was
clearly evident from some of the responses in Tables 5.60 to 5.71. This section thus highlights the most
important of these cases and establishes a basis for the revision of the contents of Table 5.59.

(a) Supply/cost of materials: As already noted, the *supply* of materials was regarded as being a problem
The label for the problem has thus been altered in Table 5.82 to read 'cost of materials'. Table 5.62
contains three responses where respondents described the problem to be that they lacked capital to finan
the purchase of materials. These responses would be more appropriately described by the label 'access t
loans/security'. Consequently, the original count of 39 for 'access to loans/security' has been increased
42 and the 'cost of materials' count reduced from 25 to 22.

Although no adjustment has been made in respect of the following, it must be noted that 10 respondents
found this to be a problem because black contractors could not obtain discounts. This could represent
increased pressure on the amount of bridging finance required (which could further increase the counts
the 'cash flow/slow payment' and/or 'access to loans/security' problems).

(b) Shortage of skilled labour: As was noted in the comments to Table 5.62, 8 responses referred to th
inability of employees to work without supervision. This is not implied by the label for the question
(shortage of skilled labour) and therefore suggests that the list of problems should have included a furth
label (perhaps - 'cost of supervision') to attract these responses. Consequently, a new label has been add
to Table 5.82 - 'cost of supervision' - and these 8 responses recorded against it. The effect of this is tha
the original count of 23 for 'shortage of skilled labour' has been reduced to 15.

(c) Preference for white contractors: About one-third (7) of the responses in Table 5.63 suggested that
real problem was that Township Councils allowed Whites to monopolise work in the townships. The lab
to the problem implies that white contractors are preferred by clients because they are more capable.
Indeed, the majority of respondents correctly interpreted it as such. Thus, the list of problem labels sho
have included an additional item, perhaps - 'Township Councils/White monopoly over township work'.
This label has been added to Table 5.82. The effect of this is that the original 'preference for white
contractors' count of 23 has been reduced to 16 and 7 counts have been allotted to the new label.

(d) Lack of technical skills: The intention was for the original label to mean a lack of technical skills on the part of the firm's employees. However, the responses in Table 5.66 indicated that, of the 18, 10 applied to the lack of technical skills of the *owner* and 8 applied to employees. This distinction should have been made in the descriptive labels for the problems. Hence, the original label has been replaced in Table 5.82 with two new labels - 'lack of technical skills (firm owner)' and 'lack of technical skills (firm's employees)'. The original count of 18 responses has accordingly been split and allocated against the new labels.

(e) Reliability of labour: Ten (10) responses in Table 5.67 came from respondents who interpreted 'reliability' to mean 'the ability to work without supervision'. Thus, these 10 responses have been transferred to the new label, 'cost of supervision', bringing its total up to 18 and reducing the count of the original 'reliability of labour' label to 7.

(f) Access to land: The comments on Table 5.69 highlighted the fact that at least 7 responses (if not all 15) connected the problem of access to land with the activities of Township Councils. Thus, 7 of these responses have been transferred to the new 'Township Councils/White monopoly over township work' label, bringing its new total up to 14 and reducing the original 'access to land' count from 15 to 8.

(g) Competition: As noted under Table 5.68, one respondent interpreted 'competition' to be a problem because of the control that Township Councils have over the flow of work in the market. Therefore, one count has been transferred to the new 'Township Councils/White monopoly over township work' label, bringing its new total up to 15 and reducing the original 'competition' count from 16 to 15.

These amendments have been executed and the new Table 5.82 now supersedes Table 5.59. Table 5.82 contains a column headed 'type of problem'. The abbreviated labels in this column refer to the categories (Labour; Finance; Production/Management; Legislation/Government; and Markets) under which the problems were grouped in Module L of the original questionnaire.

The results presented in Table 5.59 have been subsequently grouped according to problem types. The result of this exercise is presented in Figure 5.1 below. The purpose of this exercise was to eliminate the possibility of misinterpreting the results in Table 5.82. This could easily occur. For example, it is clear that 'access to loans/security' was the most common serious problem. However, this is a problem seen in isolation. It therefore cannot be taken to mean that *financial problems* were more common than all other problems. This can only be established by grouping the problems according to the type of problem, and observing which *type* was most common.

FIVE PROBLEMS WHICH WERE RANKED '4' OR '5'					
ORIGINAL ORDER OF FREQUENCY (see Table 5.59)	Count	Type of problem	REVISED ORDER OF FREQUENCY	Cou
Access to loans/security	39	Finance	Access to loans/security	
Cash flow/slow payment	30	Management	Cash flow/slow payment	
Supply/cost of materials	25	Labour	Cost of labour	
Cost of labour	24	Management	Cost of materials	
Shortage of skilled labour	23	Labour	Cost of supervision	
Preference for white contractors	23	Market	Preference for white contractors	
Lack of technical skills (firm's employees)	18	Market	Competition	
Reliability of labour	17	Labour	Shortage of skilled labour	
Competition	16	Market	Township Councils/White monopoly over township work	
Access to land	15	Management	Lack of technical skills (firm owner)	
Relationship with main contractor	10	Labour	Relationship with main contractor	
Cost of/access to equipment	9	Management	Cost of/access to equipment	
Worker action	8	Labour	Lack of technical skills (firm's employees)	
Tax legislation	8	Labour	Worker action	
Working out total costs	7	Legislation	Tax legislation	
Tender/negotiation procedures	6	Market	Access to land	
Accounting/reconciliation	5	Management	Working out total costs	
Repayment of loans	4	Labour	Reliability of labour	
Erratic inflow of work	4	Management	Tender/negotiation procedures	
Industrial Councils/labour legislation	3	Management	Accounting/reconciliation	
Relationship with sub-contractors	2	Finance	Repayment of loans	
Interest rate issues	2	Market	Erratic inflow of work	
Cost of supervision	N/A	Legislation	Industrial Councils/labour legislation	
Township Councils/White monopoly over township work	N/A	Labour	Relationship with sub-contractors	
Lack of technical skills (firm owner)	N/A	Finance	Interest rate issues	
Other legislation	0	Legislation	Other legislation	
Column total	298			

N.B. The shaded portion of the table is a reproduction of the data presented in Table 5.59

Types of problems : These are groups of problem types as presented in question L_1 of the questionnaire.

Table 5.82 - Serious Problems (Revision of Table 5.59)

The allocation of specific problem labels to the various categories could, of course, be challenged. Some readers might disagree with the totals calculated for each problem type, but a principle must nevertheless accepted - or rejected :- that problem types (which consist of various individual problems), rather than isolated problems, are more appropriate indicators of where problems lie. This will be clear from Figure 5.1 which graphically represents the total counts for each problem type. These counts were abstracted from Table 5.82 as follows: Finance (42+4+2=48); Production/Management (30+22+10+9+7+6+5=89); Labour (24+18+15+10+8+8+7+2=92); Legislation/government (8+3=11); Markets (16+15+15+8+4=58). Figure 5.1 shows these totals as percentages of the total 29

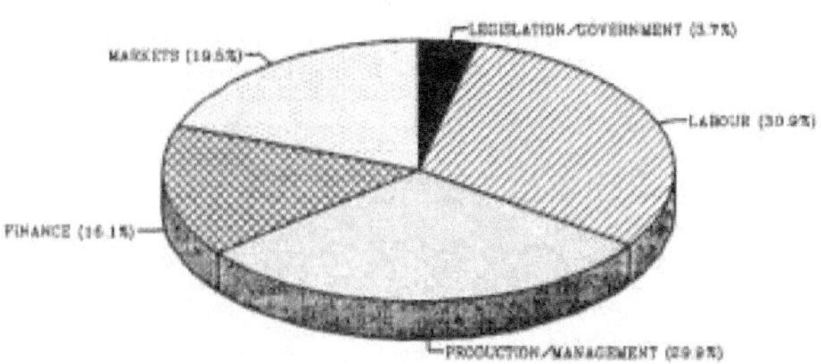

Figure 5.1 - Grouping of Problems in Table 5.82 by Type of Problem

---

5.12.4 Problems Underlying the Apparent Problems

The purpose of identifying the nature and frequency of serious problems was to establish a basis for the formulation of strategies aimed at alleviating these problems. However, caution must be exercised in drawing conclusions. These should only be drawn after a thorough understanding has been gained of the inter-relationship between the various problems. The problem is seldom what it appears to be and, as will be argued in Figures 5.2 to 5.5 below, is probably the result of underlying problems. These, in turn, are the results of further underlying problems, and so on. By hypothesising about the underlying problems, it is possible to identify likely root problems - the real causes of the apparent problems.

The following discussion probes the underlying problems for the four most common serious problems, as determined from the data in Table 5.82. Some of the underlying problems presented in the following four figures were specifically mentioned by the respondents when asked to explain the nature of the problems (see section 5.12.2). However, in most cases the author has hypothesised and included underlying problems which were generally suggested by the survey analysis. Thus, the flow charts are not definitive. Rather, it is intended that they will illustrate that problems should not be seen in isolation, but should be carefully evaluated. The formulation of policies aimed at alleviating the problems of black SSCEs should therefore target the root problems, rather than the apparent problems. Figures 5.2 to 5.5 all follow the same structure - the apparent problem (as perceived by the survey respondents) is positioned at the top of a pyramid of problems. The pyramid is divided into strata or tiers, where the lowest tier represents the *probable root causes* of the problems in the second tier, and so on. Thus, it is argued that the apparent problem is essentially a result of the root causes.

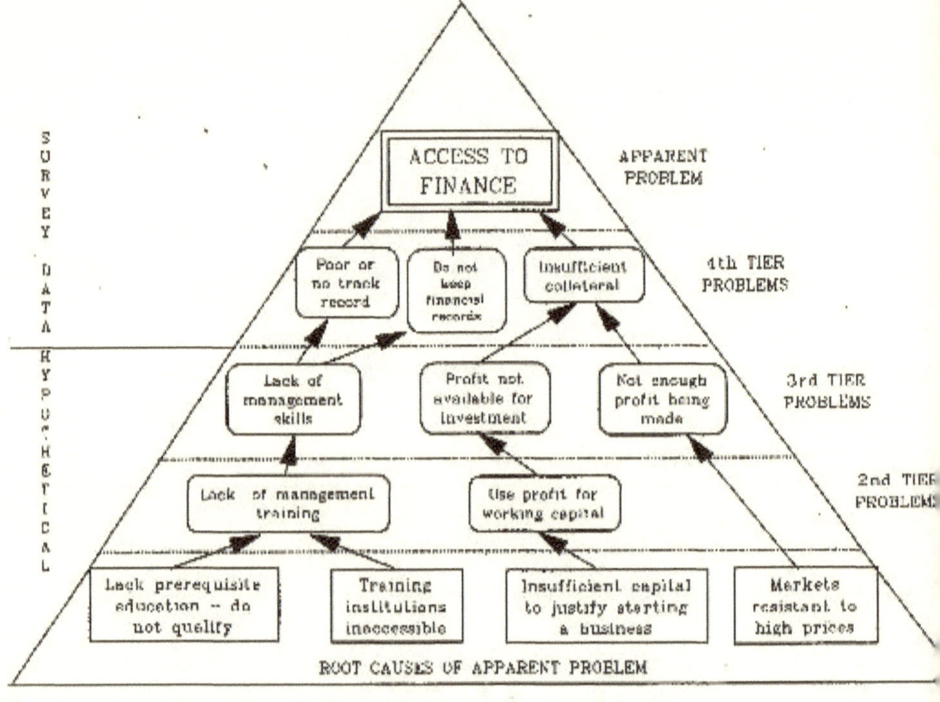

Figure 5.2 - The 'Access to Loans/Security' Problem Pyramid

From Figure 5.2, it can be concluded that lack of access to finance is not a problem, but a symptom. Tl problem is lack of collateral, or not having the business records required for loan applications. But, lacl collateral is not really the problem. Rather, the failure to have accumulated collateral is the problem. Failure to have accumulated sufficient collateral is the result of the inability to have made sufficient pro: or of neglecting to invest this profit. Failure to have made sufficient profit may be the result of poor financial management skills (caused by lack of education or training) or it could be that the market is resistant to high costs and offers no potential for profit to be made.

Thus, the probable root causes of the 'access to loans/security' problem are:

(i)   the *nature of the market* in which black SSCEs operate
(ii)   that the *business started with no capital and no prospects of obtaining it*
(iii)   the *low level of basic education* which *prevents access to certain training institutions*
(iv)   that the *management training institutions are unknown, too expensive, or remote*

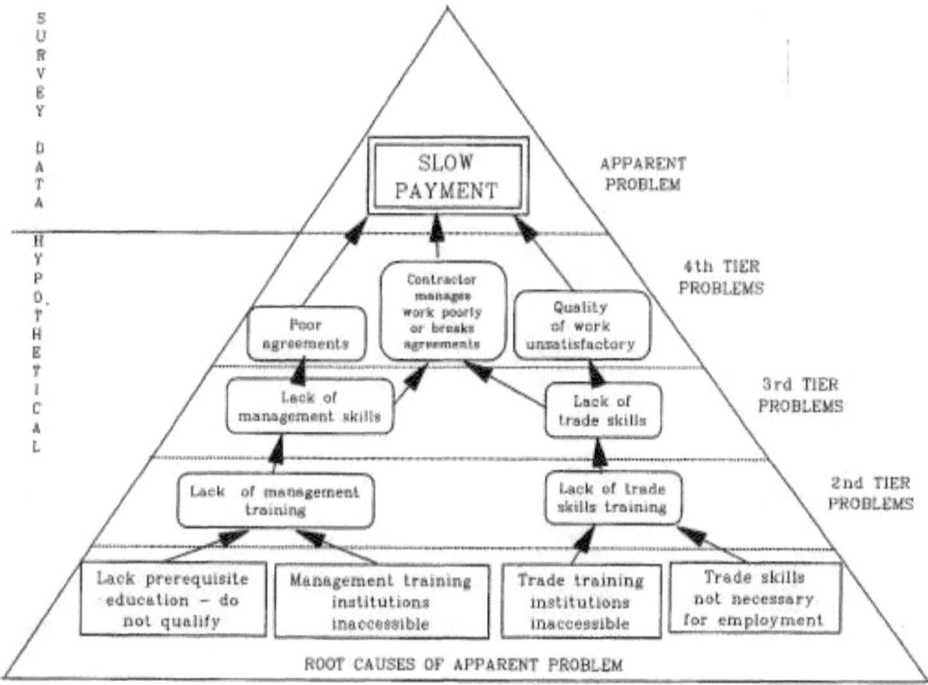

Figure 5.3 - The 'Cash Flow/Slow Payment' Problem Pyramid

The tendency for clients to pay late, or not at all, could be the result of poor agreements regarding payment, or could indicate that clients are dissatisfied with the quality of the work. However, poor agreements reflect a lack of management sophistication, which, in turn, might reflect a lack of management training. Poor quality work is the result of a lack of trade skills, or of poor supervision. The lack of trade skills among employees might be due to the inaccessibility of trade skills training institutions, but indeed, might also be the result of the willingness of contractors to employ insufficiently skilled labour.

Thus, the probable root causes of the 'cash flow/slow payment' problem are:

(i)   the *low level of basic education* which *prevents access to certain training institutions*

(ii)   that the *management training institutions are unknown, too expensive, or remote*

(iii)   that the *trade skills training institutions are unknown, too expensive, or remote*

(iv)   that *contractors are prepared to employ insufficiently skilled workers*

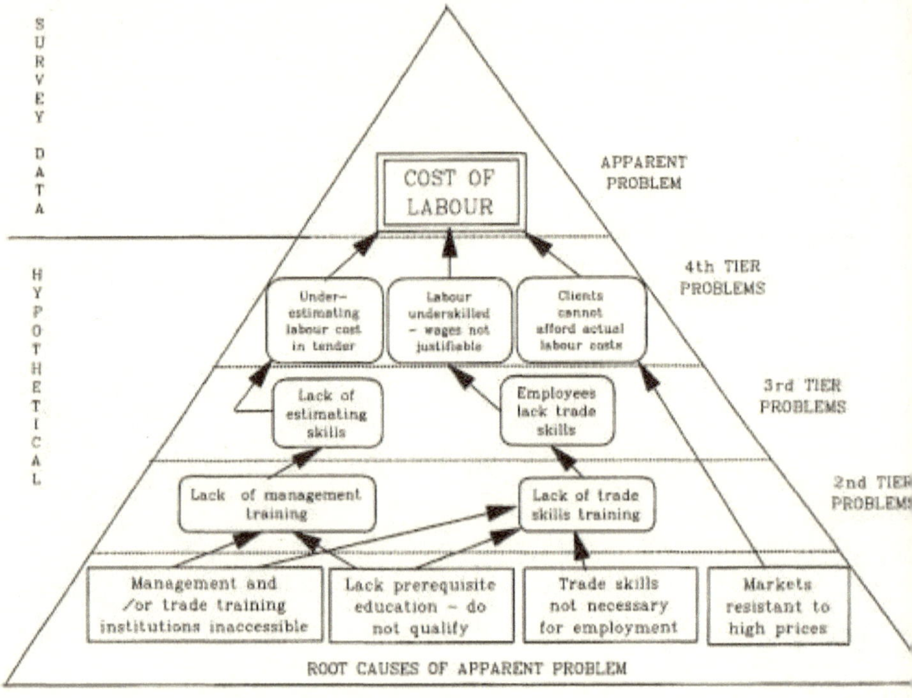

Figure 5.4 - The 'Cost of Labour' Problem Pyramid

Labour costs might appear to be excessively high if workers were unable to produce, for a given wage,
skills that the contractor would expect. In this case the underlying problems are that the contractor is
prepared to employ unqualified workers, and that workers experience difficulties entering the trade
schools/colleges. However, contractors might estimate labour productivity inaccurately at tender stage. If
the actual cost of labour exceeded the estimate, labour costs could appear to be excessive. The problem
thus one of a lack of management skills. It is, however, possible that clients exert pressure on contracto
build at prices that result in losses, hence labour costs appear to be too high.

Thus, the probable root causes of the 'cost of labour' problem are:

(i)   the *low level of basic education* which *prevents access to certain training institutions*
(ii)   that the *management training institutions are unknown, too expensive, or remote*
(iii)   that the *trade skills training institutions are unknown, too expensive, or remote*
(iv)   that *contractors are prepared to employ insufficiently skilled workers*
(v)   the *nature of the market* in which black SSCEs operate

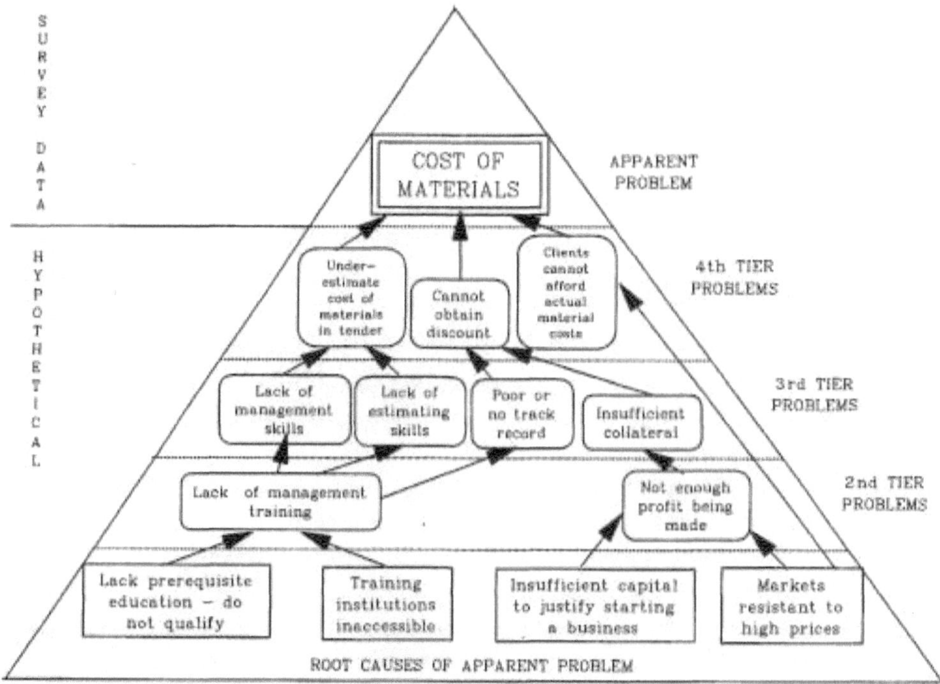

Figure 5.5 - The 'Cost of Materials' Problem Pyramid

The cost of materials could appear to be excessive if clients resist high prices and contractors must choose between making a loss on materials, or losing the contract. Alternatively, contractors who cannot obtain discounts might be perceived to be making unreasonably low profits compared with those who do secure discounts. However, it is also possible that the inability to estimate the accurate cost of materials at tender stage might result in the perception that materials costs are too high.

The inability to secure credit and/or discounts reflects the lack of confidence of the materials supplier in the contractor's ability to repay. This may be the result of lack of collateral, which is related to the inability to make and invest profits. Alternatively, lack of access to discounts might be the result of a poor track record, which in turn, is related to deficient management skills. Thus, the probable root causes of the 'cost of materials' problem are:

(i)   the *low level of basic education* which *prevents access to certain training institutions*

(ii)   that the *management training institutions are unknown, too expensive, or remote*

(iii) the *nature of the market* in which black SSCEs operate

(iv)   that the *business started with no capital and no prospects of obtaining it*

5.12.5 Summary of the Analysis of Underlying Problems

It is clear that the four apparent problems discussed in the previous section all stem from typical and related underlying problems. Perhaps the most important of the root problems is that the black residential housing market is fundamentally different from the more formal markets outside of the black areas. The relative lack of sophistication and affluence of the black markets results in contractors operating on a hand-to-mouth basis, with little prospect of achieving substantial growth.

These markets are also less demanding of formal skills, hence both trade skills and management skills tend to be underdeveloped. Not only are the markets generally tolerant of skills deficiencies, but so are contractors. Given that existing skills levels are low, it is more likely that contractors are forced to, rather than choose to, employ under-skilled workers. However, skills deficiencies are not only a result of the market's tolerance of these deficiencies. Low levels of education result in some individuals being excluded from training courses. Other individuals might qualify for training courses, but cannot afford to enrol, do not operate close enough to a training institution, or are unaware of what training courses are available.

Another root problem appeared to be that certain contractors establish their firms in a vacuum. These individuals: neglect to ensure that they have obtained sufficient *appropriate* experience; start up with insufficient capital and little or no chance of accumulating capital; and neglect to determine whether or not the market can provide them with sufficient work to justify starting a business.

CHAPTER 6

FINDINGS OF A COMPOSITE CASE STUDY OF SUCCESSFUL CONTRACTORS

In the previous chapter the problems experienced by SSCEs were identified and analysed. A logical response from a development perspective would be to address those problems. But, would this result in a general improvement of the 'successfulness' of SSCEs? As noted in section 2.6.4, there is a real possibility that it would not. This raises the issue of : what gives rise to success? Since this cannot readily be gleaned from the survey data, it was decided to interview a small sample of apparently successful contractors with the purpose of gaining an understanding of the factors of success.

This chapter reports the insights gained in interviews with three successful black contractors. These interviews were conducted in the Western Cape, rather than in the two regions from which the survey (analysed in Chapter 5) sample was taken. It was felt that this would produce reasonably representative data, since the Western Cape market is smaller and therefore more difficult to penetrate than the Johannesburg and Durban markets (that is, success would probably be more difficult to achieve in the Western Cape).

The interviewees were selected on the grounds that they had attained success as contractors and had experienced steady growth in their businesses. The interviews were conducted on an open-ended basis, taking the form of a discussion. The author guided this discussion, which was confined to the issues concerning the problems experienced by the interviewee and how these had been overcome. In the following discussion, the data from the interviews have been integrated, with the purpose of establishing generic factors of success, rather than recounting each interviewee's experiences. No references could be used in this chapter, since all three interviewees made anonymity a condition of their cooperation.

6.1 FACTORS OF SUCCESS

This section presents the author's insights into what contributes towards the success of black contractors. Common threads from all of the interviews were woven together. Factors which were perceived to contribute towards success were identified and categorised. These are presented under the headings below.

6.1.1 Personal Attributes

Success in any field is usually the result of personal endeavour. The discussion in Chapter 5 left no doubt

166

that black SSCEs face formidable difficulties in trying to establish their firms and, indeed, these difficulti
often remain unresolved. However, the individual has the choice between regarding these difficulties as
problems, or accepting them as challenges. The individual who can *turn a problem into a challenge* and
can tackle it constructively, considerably enhances his chances of success. All of the interviewees were
found to be confident individuals who attributed their success, at least in part, to having been able to tack
rather than avoid their problems.

6.1.2   Adaptability

A key factor of success is the ability to *recognise when a strategy has not worked* and, in the light of thi
to *formulate new strategies*. This applies at all stages of the individual's evolution. For example, if the ty
of experience available to an apprentice from a particular employer was inadequate, a change of employes
might solve the problem. Similarly, if the type of client the firm usually contracts with offers little scope
for profits to be made, then a change in type of client might solve the problem. The individual interviewe
had all operated in this manner from the outset of their contracting careers, always trying a new tactic
where an old one was not working. Success was, however, often not immediate, but only followed after
several such changes in direction had been made.

6.1.3   Marketing Skills

The *ability of the individual to market himself, as well as his firm*, was found to be another common
factor of success. Whether this involved obtaining a loan, securing sites, attracting clients, or establishing
materials credit lines, the approach of the interviewees was similar. All reported that many obstacles were
overcome through the ability to present themselves professionally and confidently.

A fundamental aspect to this was the *ability to know, or to establish, what was required or expected by*
party to whom the approach was being made. For example, the inclusion of a résumé, the presentation
banking records, the formulation of counter arguments, the presentation of proposed programmes, cash
flow estimates, *etc.*, are often essential components of a sound marketing strategy. Thus, the chances of
success are enhanced by the ability of the individual to differentiate between marketing techniques that w
and those that do not - and then to *discard ineffectual techniques and develop effectual tactics*.

6.1.4   Diversity of Skills

All of the interviewees were found to have effectively created bigger markets for themselves by *diversify their operations*. This either took the form of *personally acquiring additional trade skills*, or of *entering into partnership* with an individual who possessed skills not yet available to the firm. The survey data in Chapter 5, as well as the literature, suggested that black SSCEs generally resisted the notion of entering into partnerships. One of the interviewees in this case study, a sub-contractor, reported feeling this way. However, the other two interviewees attributed a great deal of their success to having expanded their firms' capabilities by entering into partnerships. The key to success was the *identification of what skills were lacking and the acquisition of those particular skills in a partner*. Thus, success resulted through broadening the capabilities of the firm, rather than through expanding it by employing more of the same skills it already possessed.

6.1.5   Controlled Growth

Another factor of success was the ability to *control the pace of growth*. The interviewees were clearly all aware of the dangers of growing too quickly and partly attributed their success to knowing when to turn work away. This was largely the result of experience and the evaluation of the firms previous activities. Success is therefore not guaranteed by growth *per se*, but by understanding the firm's limitations and operating within these.

It was also found that a *gradual accumulation of resources* (including capital) was important. Investing too much capital in equipment, vehicles, *etc.* without having the inflow of work to justify the expense, was reported as being likely to cause cash flow problems, or bankruptcy.

6.1.6   Evaluating and Understanding the Market

All of the interviewees indicated that they had entered the market only after first establishing that a need existed for their services, that is, they *recognised a niche in the market*. Once in the market, they had monitored it carefully, by *continuing to evaluate the extent to which their services were needed*. As soon as it became evident that the market was saturated, the owners re-evaluated their positions, identified new niche markets, and endeavoured to establish themselves in these. In two cases, the interviewees had expended their operations into markets within a 400 kilometre radius of their bases.

Another aspect to this was the tendency to exploit the fact that they were black, by identifying and *concentrating on markets where clients would prefer to employ black contractors*. Other such insight were also found to result in success - for example, different marketing strategies were devised for different markets.

6.1.7   Management and Organisational Ability

None of the interviewees had received formal management training. It emerged that *management experience*, rather than education, was vitally important to the achievement of success. All of the interviewees were capable of *planning individual projects as well as the overall activities of the busine* They stressed that it was not only important to be able to identify what resources were required for a project, but also to be able to organise and control the production process. One of the interviewees knew nothing about building when he was awarded 650 sites to develop. His strength was his ability to plan th project on paper, and then co-ordinate the inputs of sub-contractors. Thus, it appeared that *project plann and management were more important to success than simply knowing how to build.*

The planning and organisation of projects need not be confined to the building operations. Indeed, one interviewee noted that he extended his organisational function to include liaising with the building societ on behalf of his prospective clients, in order to facilitate their obtaining mortgage bonds.

6.1.8   Skills Training and Experience

Skills training and experience are more important for the sub-contractor than for the contractor/developer. The sub-contractor is usually his own leading artisan, as well as his own foreman. The inefficient operation of a site due to the lack of trade skills or supervision does not contribute to success or growth. However, the general contractor/developer can employ these skills via the sub-contractor.

PART C

CONCLUSIONS AND RECOMMENDATIONS

CHAPTER 7

CONCLUSIONS AND RECOMMENDATIONS

The purpose of this chapter is threefold. Firstly, a comparison is made between the findings of the empirical survey (Chapter 5) and the literature review (Chapter 2). The survey findings largely mirrored the findings presented in the literature review and the similarity between the two was highlighted throughout Chapter 5. The comparison thus includes only the major areas of difference between these two chapters. These are presented in section 7.1 below. Secondly, conclusions, based on the insights gained by the author from the analysis of the survey data, are made. Finally, recommendations are made on how the development of SSCEs could be stimulated and subjects for further research are suggested.

7.1 SUMMARY OF DIFFERENCES BETWEEN LITERATURE AND SURVEY FINDINGS

(a) Entry into self-employment via the 'trade route': In section 2.4.1 (a) it was noted from the ILO (1987) literature that most SSCEs originate from within the construction industry (via the 'trade' and 'management' routes) and from other industry and commerce (via the 'commercial' route). It was suggested that SSCE owners who entered via the 'trade route' had usually progressed through the employment ranks, that is, their experience would have included employment in unskilled, semi-skilled, skilled, and supervisory positions. Indeed, it was found (see section 5.4.1) that most firm owners had previously been building industry employees, and had thus entered via the 'trade route'. But it was clear that most *had not progressed through the ranks of employment*. Most had only had experience in one capacity, and the duration of this experience had generally been short.

Further, the ILO (1987) literature indicated that 'trade route' entrants would generally be well acquainted with the site-based activities (trade skills and management of site operations). However, this is unlikely to have been the case for many of the survey respondents, since one-third of them indicated that they required further training in trade skills (see section 5.4.4). This is clearly a consequence of the truncated nature of their previous building industry experience as employees.

It was also noted in section 2.5.1 (b) that a consistent weakness of these 'trade route' entrants was their inability to estimate and manage costs. However, in section 5.9.3 of the survey analysis, it was found that over half of the respondents did regard the calculation of tender amounts as problematic. This may, of course, have been the opinion of the firm owners, but the finding (see section 5.7.4) that about half of the respondents were unable to calculate profits accurately suggests that this is, indeed, a problem area - but one that the respondents were unaware of. A further aspect to this point is the fact that negotiation was

171

found to be the most common method of arriving at a tender amount (see section 5.7.3). This might implicitly indicate that it is unnecessary for these firms to do more than establish how much their clients are prepared to pay, rather than calculate what the work would cost.

(b) Potential for growth: The discussion in section 2.4.4 noted that some SSCEs might choose to remain small, expanding only to the point where the owner achieves a satisfactory personal income. It was found, however, (see section 5.8.1) that *the majority of firms wished to expand.* It was also suggested in section 2.4.4 that firms might choose to confine their activities to small catchment zones which provided a steady supply of new work as well as maintenance of existing buildings. This was not confirmed in the survey findings. What did emerge was that there was a general desire to cease doing alterations and additions work (see section 5.8.3), rather than an acceptance that this type of contracting represented a steady source of income.

It must also be noted that two types of growth were identified in section 2.4.4 - 'lateral expansion' (doing more contracts simultaneously) and 'upward movement' (doing larger contracts). Although the firms generally planned to expand, it appears that they would be unable to do so - neither by lateral expansion, nor by upward movement - without achieving substantial improvements in their current capabilities and capacities. This is said in the light of the finding that the majority reported that they could not undertake more than three contracts simultaneously (see section 5.9.6), and that, on average, the firms had completed 15 contracts per year (1991 only - see section 5.7.4). It will also be remembered that a turnover of 25 houses per month was considered the minimum before it became viable to enter the low-income housing market (see section 2.5.2 (f)). A more serious problem is that if housing densities increase in future as expected, SSCEs will have to be capable of building the structures which result from higher densities (for example, two- and three-storey walk-up apartments). In other words, upward movement might be essential.

(c) Labour-and material sub-contractors: In section 2.4.2, it was suggested that 'labour-and-material sub-contractors' were *specialist* sub-contractors (plumbers, electricians, *etc.*). An important fact emerged from the survey data - that many non-specialist sub-contractors indicated that they had bought materials on their first and most recent contracts. This group accounted for 30 of the 96 respondents. Since they did not identify themselves as 'general contractors', it is evident that a category of contractors exits, who operate in a sub-contracting capacity, but usually provide their own materials.

(d) Reasons for non-registration with Industrial Councils: The reasons given in section 2.5.2 (a) for why firms tended not to register with the Industrial Councils included: the high cost of registration; that there was no obvious benefit; and that the Councils were too restrictive. The survey findings confirmed these reasons, but also revealed that there were respondents who *had never heard of Industrial Councils.* In addition, there was found to be a *perception that small firms did not qualify for registration* (see section 5.11.4).

(e) The effect of BIFSA by-laws on SSCEs: In section 2.5.2 (a) it was noted that non-membership of
BIFSA had, until 1988, had the effect of denying black SSCEs access to the formal building industry.
Indeed, this is true, but the responses presented in Table 5.65 suggested that another factor was responsible
for this - that *clients preferred white contractors*. This factor suggests that, even if black contractors had
been BIFSA members, they would still have experienced problems trying to establish themselves in the
formal 'white markets'.

(f) Materials supply problems: It was noted in section 2.5.2 (f), that the erratic and unreliable supply of
materials to sites was a common problem for SSCEs. There was no evidence of this in the survey findings.
This can be established from Table 5.62, where the responses to the question regarding the nature of the
'supply/cost of materials' problem, all referred to *cost problems*. Indeed, no responses made reference to
supply problems.

(g) Township Councils: No references were found in the literature which indicated that Township Councils
were a source of problems for SSCEs. However, certain responses to the open-ended questions (see Tables
5.65, 5.68 and 5.69) gave the impression that the *Township Councils effectively control access to land in*
the townships. In the opinion of the respondents, the Township Councils cause and perpetuate the 'problem'
of white contractors monopolising township development work. In general, the respondents did not appear
to believe that they should be awarded contracts for the full developments, but felt that the Township
Councils did not do enough to ensure that they got their fair share of the sites.

In this regard, it must be noted that a statement was made in section 2.5.2 (b) that the problem of access to
land was rooted in Government land allocation policies. Indeed, this is true, in so far as it affects whether
or not land is allocated for development. But SSCEs experience the problem further down the line - at the
point where the Township Council is in a position to influence who actually gets awarded the contracts.

There are relatively few widely accessible detailed studies of South African SSCEs. There is thus a danger
that, due to their absence from the international literature, the specific problems which result because of
South African circumstances might escape consideration in the formulation of development strategies. The
issues raised above are therefore important because they correct or refine points made about SSCEs in the
literature.

7.2   CONCLUSIONS

Before proceeding to the main conclusions, some general, but important observations must be noted.
Notwithstanding the fact that the survey sample was intended to consist of 'small' (less than 20 employees)
contractors, the majority of firms were 'micro' contractors (less than 5, or no employees). This points to an
important problem area in that it suggests there are indeed very few 'small' contractors, if number of
employees is taken as the criterion. In future studies it might be more appropriate to define 'small' in terms
of alternative criteria such as turnover or capital.

Further, the majority were informal enterprises and their activities are unlikely to be reflected in the
national accounts. The Industrial Council system is unpopular, sometimes even unknown, and since non-
compliance with its agreements is illegal, it engenders informality. Unless it is restructured, there is little
likelihood that the clear demarcation between the informal and formal sectors will change.

The problems that black SSCEs identify are more commonly symptoms of related underlying problems.
Therefore, this conclusion focuses on these root causes rather than on the symptoms. Many of the
suggestions made by survey respondents as to what would improve their problems suggest that they lacked
basic entrepreneurial instincts. This was reflected in the general attitude that the rules should be changed for
black SSCEs, rather than that they should become more capable of complying with them.

The thesis of this study is that the potential for black SSCEs to contribute significantly to the output of
housing in the construction sector is limited by the internal constraints of skills-, knowledge- and
experience-deficiencies, as well as by external factors and the ineffectiveness of the existing
training/support/development infrastructure. This study validated the thesis, as is evident from the following
test of the hypothesis, which was given in section 1.3 as follows:

> **That, because of internal and external constraints and the ineffectiveness of the training/
> support/development infrastructure, black SSCEs do not currently have the capacity to play a
> significant role in formal housing development.**

It was found that black SSCEs are indeed constrained by internal and external factors as well as by
deficiencies in the scope and accessibility of the existing training/support/development infrastructure. These
points are elucidated below in the form of the following conclusions.

7.2.1   Major Internal Constraints

(a) Insufficient preparation for a career in contracting: Building contracting appears to be a 'career of last

resort' for a large number of the survey respondents, many of whom had clearly *not specifically planned*
become contractors and therefore would not have taken appropriate steps such as: (i) having obtained the
necessary theoretical knowledge in management techniques; (ii) having obtained sufficient practical
experience in both trade skills and management skills; (iii) ensuring that they had accumulated sufficient
capital to tide themselves over the start-up period; and (iv) identifying that markets existed which were
sufficiently large to sustain the firm; *etc.*.

The majority of respondents had: insufficient or no theoretical knowledge of management and planning
techniques; insufficient and unvaried practical experience; insufficient capital at start-up; and operated in
highly competitive, saturated market. In addition, a great demand existed for further training - in trade
skills, production management, and business management. There was little evidence to suggest that the
majority of respondents had obtained sufficient training in these fields. If this were to be remedied, they
would only be able to participate in on-the-job training or in training programs of a part-time or
correspondence nature, since the firm owner generally performs the joint function of skilled worker and
foreman/supervisor. Further, it was clear that the demand was preferably for training that would result in
the award of a qualification (certificate, diploma, degree, *etc.*).

(b) Growth history and potential extremely limited: The growth of black SSCEs, if it occurred at all,
tended to be slow. Types of client and types of building constructed tended to remain unchanged over the
life of the firms - the majority of contractors were employed by individuals to build or alter houses.
Monthly turnovers (while contracting) only increased marginally over the life of the firms. These turnov
are still generally low - mostly less than R 6 000 per month. The numbers of firms that made losses
decreased only slightly over the life of the firms. However, profits were relatively low and the amount o
profit made per contract tended to reduce over the life of the firm. The majority of contractors achieved
only slight improvements in the value of their capital over their firms' life spans. However, the majority
black SSCEs wish to expand their firms. The potential for this to be achieved is severely limited due to
following five factors.

(i) The reluctance to enter into partnerships. The tendency towards sole-ownership largely represents a f
of losing control over the financial management of the business. A serious consequence of this tendency
the loss of the benefits that would accrue to the firm from the synergy of skills of a number of partners
well-balanced partnership. Thus, the reluctance to enter into partnerships is a serious constraint to latera
growth, given that the majority of firms could not undertake more than two to three simultaneous contra
with their current management structures.

(ii) Black 'general contractors' tend not to sub-contract work out to others, preferring to retain control b
doing it themselves. This, however, has the effect of diminishing the size of the sub-contracting market.

(iii) The lack of access to bulk-buying facilities may appear to be an external constraint, but it derives from the internal constraint of lack of creditworthiness.

(iv) The general inability to raise collateral does not appear to diminish as the firm grows older, suggesting that many usually engage in 'hand-to mouth contracting'. The sluggish growth in capital over the life of the firms, as well as the fact that ability to raise collateral does not improve, must indicate that what little profits are made are not reinvested in the business.

(v) Firms were unable to establish themselves in more desirable niche markets. The majority who usually undertook alterations and additions contracts expressed a desire to move away from this type of contracting and towards the construction of new buildings. Thus, alterations and additions contracting appears to be a relatively easy way to enter the market. However, it appears that many find it difficult to progress to contracts for new buildings. Sub-contractors expressed a desire to become general contractors, which appears to be a reaction to exploitation.

7.2.2 Major External Constraint

(a) Nature of the market: The majority of black SSCEs are locked into the black residential housing market. The reticence of the financial institutions to extend mortgage finance to black home-owners results in a decrease in the size of the market and fuels competition among contractors. Other markets will be difficult to penetrate, especially those where the competition is White. Black residential markets differ from those in which small white contractors operate in four key respects.

(i) Black residential markets are more resistant to high, and rising, building costs, evidenced by the fact that negotiation and take-it-or-leave-it offers are more prevalent than competitive tenders. Further evidence of this is that: contractors consider labour costs and material costs to be too high, probably reflecting clients' resistance to paying more than what they consider to be an acceptable rate per square metre; and clients often suspend payments, causing contractors to leave sites prematurely. This aggravates the poor reputation that black SSCEs generally have.

(ii) The markets are more informal, involving smaller and less sophisticated buildings. They are thus more forgiving of deficiencies in trade skills. In addition, the markets do not appear to display any resistance to the fact that many firms are unregistered, and they are accepting of unsupervised contracting - the presence of a foreman on a site is the exception rather than the rule.

(iii) It is common practice in these markets for deposits to be paid and these substitute for formal working capital.

(iv) The markets are more politicised. This is evidenced by the powerful role played by the Township
Councils in controlling the supply of development land and the need to form lobbying groups ('contracto
associations') to interface with these bodies.

Black SSCEs wishing to break out of these markets into the white residential markets face formidable
barriers. These include:

(i) The lack of literacy in English. The majority of survey respondents did not complete their secondary
schooling and also cited English as a second language. It is therefore likely that many would experience
problems reading complex contract documents, specifications, *etc.*. It is therefore also likely that Englis
speaking clients, financiers, building inspectors, *etc.* would harbour reservations as to the general
capabilities of black SSCEs.

(ii) Low levels of education and lack of formal qualifications put these contractors at a disadvantage relat
to their white counterparts.

(iii) The payment of deposits is uncommon in the white markets. Clients' deposits are an important sour
of bridging finance to black SSCEs - who generally fail to obtain finance from formal lending institution
Thus, a shift from the informal less sophisticated markets into the formal markets would be accompanied
the need for the contractor to provide his own bridging finance.

7.2.3 Ineffectiveness of Existing Training/Support/Development Infrastructure

The support available to SSCEs from the various institutions which attempt to provide it, is severely
limited. CAs are an important link between the general SSCE fraternity and development institutions and
funding, but they represent a minority of SSCEs. This is a serious problem, both in terms of what they
themselves can provide (internal services), as well as what they can connect the SSCE with, for example
benefactors (external representation). Both the PMDIs and the NPMDIs have their strengths and
weaknesses, but the overriding consideration in terms of their effectiveness, is their relative inaccessibili
both in terms of location and entry qualifications. It thus is concluded that, collectively these organisatio
are minor sources of work, technical support, and training.

7.2.4 Limited Capacity

Recent and current outputs, expressed in terms of total numbers of contracts completed per year as well capacity to do contracts simultaneously, show that the respondents generally lack the capacity to contribute significantly towards the provision of housing. Indeed, all of the internal and external constraints noted in the previous three sections either cause, contribute to, or aggravate the capacity problem. No doubt capacity can be improved over time, and the majority of firms desire this, but it is unlikely that this will occur without a great deal of support and access to experience in more formal construction methods and markets.

The following section will address this area by suggesting possible steps which could be taken to improve access to training, support and experience.

7.3 RECOMMENDATIONS

This section contains recommendations regarding the support and promotion of South African black SSCEs. The recommendations flow from the analysis of the survey data, the conclusions, and are also influenced by the discussion in section 2.6 of the literature review. It is expected that a variety of organisations in the public and private sectors could implement these recommendations. No attempt has been made to articulate exactly who should be responsible for what, since it is the purpose of this section to only recommend action.

7.3.1 Recommendation No. 1 - Analyse and Improve the Existing Training Infrastructure

The geographical distribution of training institutions needs to be studied, since it is clear that certain regions are better catered for than others. What is needed is a better balance. This could be achieved by establishing new institutions, or introducing new courses at appropriate existing institutions. Further action should involve the identification of existing training institutions which currently cannot meet the demand for their courses and improve their capacity.

Something needs to be done about helping applicants get accepted into training programmes where they are excluded because of stringent entry requirements. This could possibly be accomplished by devising and implementing short bridging courses tailored to enable applicants to meet these requirements.

In addition to formal training programmes, there is a specific need for specially tailored part-time and/or correspondence training courses dealing with planning, marketing and management. This need could be satisfied by supporting the expansion of existing courses, or by implementing new courses through existing institutions. There is also a need for similar initiatives to be taken in respect of trade skills training.

The above recommendations all require extensive research, course design, *etc.*. The likelihood is that many will not be implemented because of this, or that they would only be implemented in the medium- to long-term. In the meanwhile, an alternative would be to support the existing programmes (especially the

NPMDIs) which currently promote the development of SSCEs. This support should be aimed at enabling these programmes to reach more SSCEs and to establish themselves nationally, where appropriate.

Contractors' associations need support. The nature of this support should be such that it makes them attractive to belong to, that is, they need to be made more representative. Once this has occurred, they will require (at least initially) support to improve their capability of providing support to their members. The associations need to offer support such as: making available simplified model documentation such as contracts and specifications, written in African languages; regularly organising and presenting mid-career training seminars; publishing periodicals which include instructional material covering trade skills, as well as management and planning skills; *etc.*.

Thus far it has been noted what needs to be provided in the support of SSCEs. If, however, all of this were accomplished, there would still be the problem of affordability. It is therefore recommended that support be provided for the establishment of bursary schemes to enable emerging and established SSCEs to enrol for degree, diploma, and certificate courses at universities, technikons, technical colleges, and other institutions. In the case of established contractors, more emphasis should be placed on facilitating access to short courses.

7.3.2   Recommendation No. 2 - Improve Access to Work and Experience

National and local government could introduce some stability into the contracting market through the annual publication of a programme of planned government and quasi-government contracts and this is recommended. It is further recommended that the introduction of tender biases be promoted. This could ensure that local or community-based contractors are ensured of involvement in government and local authority developments. In addition, it is critical that all types of contractors (from labour-only sub-contractors to developers) are included on these schemes.

Further, where private sector developers undertake developments, they should be offered incentives to include all types of black SSCEs in such developments. This should also take the form of encouraging joint ventures between capable contractors and emerging contractors - perhaps through a subsidy scheme whereby capable contractors qualify for subsidies if they enter into consortia or joint venture partnership with emerging black SSCEs.

In essence, it is important that black SSCEs gain access to on-the-job training and experience, bearing in mind that, in the short term, they would be better suited to small works and house building. However, making this recommendation, it must be noted that one of the things that has not worked to date, is the linking of the development of SSCEs to the production of housing. This was seen in the lacklustre performance of the SAHT and the retreat of the UF HUCs from conventional housing into the relatively safe haven of site-and-service. But, linking SSCEs development to housing development also seems to be one of the few strategies that could work. This paradox must, however, not be allowed to mislead us, since the reasons why SSCE development and housing production have not gone well together are often due to political factors. Township violence, theft on sites, bond boycotts, *etc.* have all taken their toll and, at a time when housing is desperately needed, the major private sector developers have withdrawn from the low-income housing market. This has had the effect of placing enormous pressure on those SSCE development institutions that produce housing, because they have basically become the major clients in the low-income housing market. Consequently, SSCEs are attracted to them because they are effectively a market. But this interferes with the more important function they could be performing - that of training SSCEs for the rest of the (currently non-existent) housing market.

The advent of political stability in South Africa will alleviate some of these problems and enhance the prospects of successfully merging SSCE development and housing production. If, at that time, all parties involved in SSCE support/development or housing production work within a co-ordinated framework, results on a scale not yet seen will be possible.

7.3.3   Recommendation No. 3 - Establish a Contractor Development Agency

Notwithstanding the ambivalent findings regarding the potential for CDAs to accomplish anything substantial in the development of SSCEs, it is recommended that such an organisation be established. This should not, however, be undertaken without a thorough analysis of why previous ventures failed.

Whatever form it ultimately took, a CDA should perform the following functions:

(i) the CDA should be able to inform contractors of all existing training and development programmes offered by existing institutions. In addition, it should be equipped to assess the training needs of contractors and to direct them to the appropriate institutions.

(ii) the CDA should be capable of designing and implementing training courses which address the immediate problems of black SSCEs, particularly those needs which are not serviced by existing programmes. These courses should be designed in consultation with the prospective trainees.

(iii) the CDA should perform the role of 'crisis counselling'. Contractors should have access to a resource whereby their immediate problems (for example, a rejected loan application) can be assessed and appropriate courses of action recommended.

(iv) the CDA should provide general counselling on legal matters. This service should be able to address issues specific to particular projects (for example, contracts) as well as to the business itself (for example, insolvency).

(v) the CDA should assist contractors who fail to obtain loans from other institutions. Types of support which should be offered include: the extension of bridging finance (short-term loans); the provision of sureties; and the extension of medium-term investment loans.

(vi) the CDA should provide technical support relating to site-based operations

(vii) the CDA should provide, or arrange for ready access to hired equipment

(viii) the CDA should facilitate access to discounts on materials, either by purchasing and stockpiling materials, or by providing guarantees which allow SSCEs to establish credit lines with suppliers

(ix) the CDA should design and offer an estimating support service. This should assist contractors to measure quantities, determine costs, assess wastage, estimate labour productivity, *etc.*.

7.4 SUBJECTS FOR FURTHER RESEARCH

This study has identified that all of the following subjects are causes of, or are related to the problems f.
by black SSCEs. They have also been probed and analysed in this study, but to the extent that the scope
the study permitted. It is thus recommended that more detailed research be undertaken in all of these are

7.4.1 Little is known of the 'informal' sector of the South African construction industry. A study
necessary which defines the sector, estimates its size and value of work done by it, identifies its
markets, *etc.*.

7.4.2 **The Labour Relations Act and the Industrial Council agreements**. Are these agreements
effective? Are they appropriate? How do the various agreements for the five main metropolitan
areas differ?

7.4.3 **Materials supply cartels**. Do they exist? What effect do they have on building costs incurred by
various types of contractors?

7.4.4 **The training infrastructure in the South African construction industry**. Are adequate faciliti
available for skills, management and entrepreneurial training, particularly with respect to small
building contractors? What should be done to create an appropriate and effective training
infrastructure.

7.4.5   **Market barriers facing the black SSCE.** How are these small contractors inhibited in their attempts at establishing themselves in "white" markets?

7.4.6   **The supply of land regarding housing developments for low-income communities.** What is the role and influence of the local authority? What do contractors have to do to obtain sites?

7.4.7   **Sub-contracting.** How widespread is this phenomenon? Does it affect quality? Should it be promoted or discouraged?

REFERENCES

JOURNAL ARTICLES, REPORTS, *etc.*

African Building Contractor (ABC) (1988a), "Land acquisition and house plans", September/October, p. 48.

ABC (1988b), "A.M. Makena home builders", November/December, p. 30.

ABC (1988c), "A history lesson in the building industry", November/December, p. 30.

ABC (1988d), "Vukoma Housing", November/December, p. 35.

ABC (1988e), "Ennerdale Builders Guild - an independent body", November/December, p. 35.

ABC (1988f), "B.A. Sello Building Construction", November/December, p. 12.

ABC (1988g), "Khaya Elihle Building Construction", November/December, p. 3.

ABC (1988h), "Vosloorus Builders & Allied Trade Association", November/December, p. 12.

ABC (1989a), "Joe Taylor leaves S.A. Housing Trust", May/June, pp. 8-12.

ABC (1989b), "The servicing of land for residential towns - the SA Housing Trust highlights some major considerations", November/December, pp. 32-33.

ABC (1989c), "NAFCOC and the Professional Builders Federation hold builders workshop", reprint of speech given at seminar (The Black Building Industry - Prospects for economic upliftment) by Linda Nyembe, PBA chairman, at Pretoria 27 June 1989, September/October, p. 17.

ABC (1989d), "Black, white joint ventures are the only solution to the housing problem", May/June, pp. 48-49.

ABC (1989e), "The role of the emerging small professional builder", reprint of speech given to the Professional Builders Federation by Joe Hlongwane, NAFCOC vice-president, May/June, pp. 40-47.

ABC (1989f), "SA Housing Trust involved in Alex development", August, p. 20.

ABC (1989g), "SA Housing Trust shows the way in creating small building companies - Jouberton project proves value of teaching black builders to run their own companies", September/October, p. 40.

ABC (1989h), "The Soweto Building Contractors Ass.", September/October, p. 22.

ABC (1990a), "Sixteen well trained Soweto builders receive certificates", January/February, pp. 12-13.

ABC (1990b), "The second AGM of the Professional Builders Federation - chairman calls for more market information", reprint of speech given at the second AGM (Theme for the day - Black Economic Empowerment) of the PBA at Johannesburg, 18 November 1989, by president Linda Nyembe, January/February, pp. 7-10.

ABC (1990c), "Training, finance and advice now available", October/November, pp. 29-30.

ABC (1990d), "Strong links forged between black and white builders' associations", October/November, p. 11.

ABC (1990e), "Palm Springs - an SAHT project", March April, pp. 20-21.

ABC (1990f), "Development programme scored in SAHT Small Builder Contest", October/November, 18.

ABC (1990g), "Housing Trust helps to train small builders", October/November, p. 25.

ABC (1991a), "Now the professional small builder can compete thanks to the new CDA scheme", April/May, p. 12.

ABC (1991b), "Very positive TABA AGM", April/May, p. 7.

ABC (1991c), "Winners of the S.A. Housing Trust's small builder of the year competition", December/January, pp. 4-6.

ABC (1991d), "A profile of the S.A. Housing Trust", September, pp. 24-25.

ABC (1991e), "Reef training centre for builders could be for you", December/January, p. 12.

Alexander, G.D. (1990), "A partnership that 'over-extended' itself but was rescued from failure", *African Building Contractor*, January/February, pp. 22-30.

Archer, S. (1988), "'The best of all possible worlds'. debating alternative economic systems in South Africa", *Sash*, December, pp. 5-12.

Bloemfontein Building Industry Development Foundation (BBIDF) (1989), "Bouaannemersontwikkelingsprogram BAO (Vlak 1) -Kursus vir klein bouaannemers - Verslag 1989", report obtained from Professor B. Verster, University of the Orange Free State.

BBIDF (1990), "Bouaannemersontwikkelingsprogram - Kursus vir klein bouaannemers - Verslag 1990", report obtained from Professor B. Verster, University of the Orange Free State.

BBIDF (1991), "Bouaannemersontwikkelingsprogram BOA (Vlak 1) -Kursus vir klein bouaannemers - Verslag 1991", report obtained from Professor B. Verster, University of the Orange Free State.

Building Industries Federation of South Africa (BIFSA) (1983), BIFSA Annual Report.

BIFSA (1984), BIFSA Annual Report.

BIFSA (1985), BIFSA Annual Report.

BIFSA (1986), BIFSA Annual Report.

BIFSA (1987), BIFSA Annual Report.

BIFSA (1988a), BIFSA Annual Report.

BIFSA (1988b), Circular 1988/11 - ADM. 201.01. Correspondence to members detailing revised by-laws referenced RMA/11b, 20 January 1988.

BIFSA (1989), BIFSA Statistical Yearbook.

BIFSA (1990), BIFSA Statistical Yearbook.

BIFSA (1992), BIFSA Statistical Yearbook.

Burgess, R. (1978), "Petty commodity housing or dweller control? A critique of John Turner's views on housing policy", *World Development*, Vol.6, No. 9, pp. 1105-1133.

Calcopietro, C.M. (1992), "Institutions supporting small-scale builders in South Africa"; unpublished paper obtained from the author, pp. 1-42.

Courier (1991), "Decade of Achievement", Vol. 9, No. 1, p. 3.

Creighton, P., (1992), "The Hodeco Operation", unpublished address given to M.B.S. (Pty) Ltd. at Downtown Inn on 25 January 1992, transcript obtained from the author.

Community and Urban Services Support Project (CUSSP) (1993), Information brochure obtained from Cape Town office of CUSSP.

Currie, L. (1971), "The exchange constraint on development : A partial solution to the problem", *Economic Journal*, 81, pp. 886-903.

Duncan, K. (1991), "An integrated approach to the economic empowerment of small builders in South Africa", article obtained from the author (excerpts from an article written in August 1991 proposing the repositioning in the market of the SA Housing Trust's Small Builders Department).

Duncan, K. (1992a), "Economic empowerment through the SA Housing Trust's new Small Builder Programme", - article obtained from the author (an article written for the Sowetan in January 1992).

Duncan, K. (1992b), "Small Builders Department", article obtained from the author (article written for SA Housing Trust's in-house magazine in February 1992).

Entrepreneurial Development (Southern Africa) (EDSA) (1990), Annual Report 1990.

EDSA (1991), Annual Report 1991.

EDSA (1992),   Annual Report 1992.

Ferreira, F.H. and Potgieter, J.F.   "The venture capital industry in south africa and the potential for participation by black
(1989),   entrepreneurs", Institute of Social and Economic Research, Development Studies Unit,
                                      Rhodes University, Grahamstown, Working Paper No. 48, pp. 53-55.

Financial Mail (1986),   "Building cartel : coming to terms", Vol. 102, No. 11, 12 December, p. 47.

Fraser, N. (1989),   "Under-quoting is a formula for disaster", *African Building Contractor*,
                                      September/October, pp. 12-16.

Griffin, C. (1990),   "Don't get caught-out by price increases", *African Building Contactor*, January/February
                                      pp. 18-19.

Hendler, P. (1993),   "Analysis, critique and strategic implications of the De Loor Report", report prepared fo
                                      Planact.

Housing in Southern Africa (HSA)   "Outgoing chairman sees '89 as a year filled with many challenges and opportunities",
(1989a),   April, pp. 9-10.

HSA (1989b),   "South Africans spend billions on additions and extensions", May, p. 12.

HSA (1989c),   "Builders urged to form closer ties with local authorities", August, p. 18.

HSA (1989d),   "Investing in the future of the country", November/December, pp. 20-23.

HSA (1989e),   "What is the SA Housing Trust?", January, pp. 13-15.

HSA (1990a),   "Whites not keen to promote Blacks", February, p. 33.

HSA (1990b),   The ANC and COSATU on the provision of new affordable homes in a post-Apartheid
                                      R.S.A", August, pp. 4-5.

HSA (1991a),   "The time for theorising and politicising is over - Hardy", February, pp. 11-13.

HSA (1991b),   "The low-cost market is going to become the dominant market - SAHT", February, p. 2

HSA (1991c),   "The era of the large building contractor is over - survey", March, p. 9.

HSA (1991d),   "Housing for all - no longer just a dream", November/December, pp. 18-19.

HSA (1991e),   "The DBSA plays a vital role in securing a better future for all", April, p. 21.

Koenigsberger, O. (1986),   Third World Housing Policies since the 1950s", *Habitat International*, Vol. 10, No. 3, p
                                      27-32.

Krafchik, W. (1991),   "Growth possibilities and constraints in the Western Cape construction industry", in
                                      Working Papers on the Economy of the Western Cape, Paper 2, pp. 1-24.

Kwazulu Training Trust (KTT)   Annual Report 1992.
(1992),

Kwazulu Finance and Investment   Annual Report 1992.
Corporation Limited (KFC) (1992),

Mabe, S. (1989),   Cited in *African Building Contractor* (1989c), "NAFCOC and the Professional Builders
                                      Federation hold builders workshop", reprint of speech given at seminar (The Black
                                      Building Industry - Prospects for economic upliftment) by Linda Nyembe, PBA chairman
                                      at Pretoria 27 June 1989, September/October, p. 17.

Merrifield, A. (1992),   "Private sector involvement in South Africa's low-income housing market since the late
                                      1980s". Unpublished report, Department of Property Development and Construction
                                      Economics, University of Natal.

Mngomezulu, S. (1988),   "Builders loans", *African Building Contractor*, September/October, p. 10.

Motlanthe, P. (1990),   "Critical needs of the emerging contractor", *African Building Contractor*,
                                      January/February, pp. 15-16.

National African Federated   Constitution - obtained from Mr. C. Petersen of NAFBI, pp. 1-13.
Chamber for the Building Industry
(NAFBI) (1991),

NAFBI (1992),

"Proposal - Establishment of Technical Aid Centres", unpublished circular - obtained from Mr. C. Petersen of NAFBI, pp. 1-3.

Nientied, P and van der Linden, J. (1985),

"Approaches to low-income housing in the third world: some comments", *International Journal of Urban and Regional Research*, 9 (3), pp. 311-328.

Nonyane, D. (1988),

"A new house needed every 40 seconds!", *African Building Contractor*, September/October, p. 2.

Padi, D. (1988a),

"Success stories", *African Building Contractor*, September/October, pp. 3-9.

Padi, D. (1988b),

"Success stories", *African Building Contractor*, November/December, pp. 2-3.

Padi, D. (1988c),

"A professional A.G.M. for a truly professional organisation", *African Building Contractor*, November/December pp. 6-7.

Padi, D. (1989a),

"Success stories", *African Building Contractor*, May/June, pp. 2-5.

Padi, D. (1989b),

"Success stories", *African Building Contractor*, August, pp. 3-6.

Padi, D. (1989c),

"Success stories", *African Building Contractor*, September/October, pp. 2-8.

Padi, D. (1990a),

"Success stories", *African Building Contractor*, January/February, p. 3.

Padi, D. (1990b),

"Success stories", *African Building Contractor*, March/April, p. 2-8.

Padi, D. (1990c),

"Success stories", *African Building Contractor*, May/June, p. 2-8.

Roukens De Lange, A. (1990),

"The Impact of Provision of Low-Cost Housing on the South African Economy", report published by the Institute for Futures Research, University of Stellenbosch, pp. 11-17.

Slingsby, M. (1986),

"Community Development Support Programmes for Housing Projects - A Problem-Solving Approach", *Habitat International*, Vol. 10, No. 3, pp. 65-71.

Tholiwe Homes (TH) (1992),

Internal reports, *etc.* obtained from Tholiwe Homes.

Turner, J.F.C. (1972),

"Housing as a verb", in *Freedom To Build*, J.F.C. Turner and R. Fichter (eds.). The Macmillan Company : New York.

Turner, J.F.C. (1978),

"Housing in three dimensions : Terms of reference for the housing question redefined", *World Development*, Vol.6, No. 9, pp. 1135-1145.

Urban Foundation (UF) (1988),

"The Urban Foundation Annual Review 1988", pp. 1-29.

UF (1989a),

"Role and Achievements", unpublished internal pamphlet, pp. 1-14.

UF (1989b),

"The Urban Foundation Annual Review 1989", pp. 1-21.

UF (1990),

"The Urban Foundation Annual Review 1990", pp. 1-39.

UF (1991),

"The Urban Foundation Annual Review 1991", pp. 1-36.

UF (1992),

"The Urban Foundation in 1992", unpublished internal pamphlet, pp. 1-9.

Van Staden, T. (1993),

"A holistic approach to SME development through grassroots community participation", *Proceedings of the 21st IAHS World Housing Congress*, Vol. 1, pp. 579-590.

Van Niekerk, C.J. du P and Hutton, R. (1993),

"Affordable 100% steel housing : A manufacturer's experience in the RSA", *Proceedings of the 21st IAHS World Housing Congress*, Vol. 1, pp. 567-568.

Watermeyer, R.B. (1993),

"Community-based construction : Mobilising communities to construct their own infrastructure", *Proceedings of the 21st IAHS World Housing Congress*, Vol. 1, pp. 603-614.

Wakely, P. (1986),

"The Devolution of Housing Production : Support and Management", *Habitat International*, Vol. 10, No. 3, pp. 53-63.

BOOKS

African National Congress (ANC) (1993),

Ready to Govern. African National Congress policy guidelines for a democratic South Africa adopted at the National Conference 28-31 May 1992.

BIFSA,

BIFSA Official Handbook (loose leaved - no identifiable publication date)

Cohen, S. (1974),
Bukharin and the Bolshevik Revolution. A Political Biography 1888-1938. Wildwood House : London.

Cohen, S. (1986),
Rethinking the Soviet Experience. Oxford University Press : New York.

Edmonds, G.A. and Miles, D.W.J. (1984)
Foundations For Change. Aspects of the construction industry in developing countries. Intermediate Technology Publications : London.

Engels, F. (1979),
The Housing Question. Progress Publishers : Moscow.

Harrington, M. (1977),
The Twilight of Capitalism. Macmillan : London.

Hendler, P. (1989),
Politics on the Home Front. South African Institute of Race Relations : Johannesburg.

International Labour Organisation (ILO) (1987).
Guide-Lines for the Development of Small-Scale Construction Enterprises. International Labour Organisation : Geneva.

Kay, G. (1975),
Development and Underdevelopment. A Marxist Analysis. Macmillan : London.

Lekachman, R. and Van Loon, B. (1981),
Capitalism for Beginners. Writers and Readers Publishing Co-op. : London.

Marx, K. (1973),
Grundrisse: Foundations of the Critique of Political Economy. Penguin : Harmondworth cited by Nove (1986a), *Marxism and "Really Existing Socialism"*. Harwood Academic Publishers : London.

Moll, P. (1990),
The Great Economic Debate. Skotaville Publishers : Johannesburg.

Nove, A. (1986a),
Marxism and "Really Existing Socialism". Harwood Academic Publishers : London.

Nove, A. (1986b),
Socialism, Economics and Development. Allen & Unwin : London.

Preobrazhensky, E.A. (1980),
The Crisis of Soviet Industrialization. Macmillan : London.

Strassmann, P.W. (1978),
Housing and Building Technology in Developing Countries. Michigan State University International Business and Economic Studies : Michigan.

University of Cape Town Careers Office (1991).
Bridging The Gap 1992 - A Guide to Practical & Educational Courses in the Western Cape. UCT Careers Office : Cape Town.

Wells, J. (1986),
The Construction Industry in Developing Countries : Alternative Strategies for Development. Croom Helm : Beckenham.

World Bank (1984).
The Construction Industry - Issues and strategies in developing countries. The World Bank : Washington.

GOVERNMENT PUBLICATIONS

Census of Construction, (1985),
See Republic of South Africa, (1985).

Republic of South Africa (1985),
"Census of Construction, 1985", Report No. 50-01-01 (1985).

Republic of South Africa (1987a),
Regulation Gazette No. 4114, Government Gazette, Vol. 265, No.10844, 31 July 1987, pp. 1-35.

Republic of South Africa (1987b),
Regulation Gazette No. 4078, Government Gazette, Vol. 262, No.10720, 24 April 1987, pp. 1-17.

Republic of South Africa (1989),
Regulation Gazette No. 4332, Government Gazette, Vol. 285, No.11766, 23 March 1989, pp. 3-4.

Republic of South Africa (1992a),
Regulation Gazette No. 4838, Government Gazette, Vol. 321, No. 113852, 20 March 1992, pp. 9-28.

Republic of South Africa (1992b),
"SA Statistics, 1992". Central Statistical Services, Pretoria.

SA Statistics (1992),
See Republic of South Africa, (1992b).

South African Housing Advisory Council (SAHAC) (1992),
"Housing in South Africa : Proposals on a policy and strategy" (The De Loor Report).

THESES

Krafchik, W. (1990),
Small-Scale Enterprises, Inward Industrialisation, and Housing : A Case Study of Subcontractors in the Cape Peninsula Low-Cost Housing Industry. Unpublished Masters thesis, School of Economics, University of Cape Town.

Rolfe, S.M. (1990),
The Stimulation of Small-Scale Enterprises in South Africa Through the Process of Subcontracting. Unpublished Honours thesis, School of Economics, University of Cape Town.

Small-Scale Enterprises, Inward Industrialisation, and Housing : A Case Study of Subcontractors in the Cape Peninsula Low-Cost Housing Industry.

The Stimulation of Small-Scale Enterprises in South Africa Through the Process of Subcontracting. Unpublished Honours thesis, School of Economics, University of Cape Town.

BIBLIOGRAPHY

African Building Contractor (ABC) (1988a), "Land acquisition and house plans", September/October, p. 48.

ABC (1988b), "A.M. Makena home builders", November/December, p. 30.

ABC (1988c), "A history lesson in the building industry", November/December, p. 30.

ABC (1988d), "Vukoma Housing", November/December, p. 35.

ABC (1988e), "Ennerdale Builders Guild - an independent body", November/December, p. 35.

ABC (1988f), "B.A. Sello Building Construction", November/December, p. 12.

ABC (1988g), "Khaya Elihle Building Construction", November/December, p. 3.

ABC (1988h), "Vosloorus Builders & Allied Trade Association", November/December, p. 12.

ABC (1988i), "Dan Psdi looks at the 'amazing gap'", November/December, p. 3.

ABC (1989a), "Joe Taylor leaves S.A. Housing Trust", May/June, pp. 8-12.

ABC (1989b), "The servicing of land for residential towns - the SA Housing Trust highlights some major considerations", November/December, pp. 32-33.

ABC (1989c), "NAFCOC and the Professional Builders Federation hold builders workshop", reprint of speech given at seminar (The Black Building Industry - Prospects for economic upliftment by Linda Nyembe, PBA chairman, at Pretoria 27 June 1989, September/October, p. 17.

ABC (1989d), "Black, white joint ventures are the only solution to the housing problem", May/June, pp 48-49.

ABC (1989e), "The role of the emerging small professional builder", reprint of speech given to the Professional Builders Federation by Joe Hlongwane, NAFCOC vice-president, May/June, pp. 40-47.

ABC (1989f), "SA Housing Trust involved in Alex development", August, p. 20.

ABC (1989g), "SA Housing Trust shows the way in creating small building companies - Jouberton project proves value of teaching black builders to run their own companies", September/October, p. 40.

ABC (1989h), "The Soweto Building Contractors Ass.", September/October, p. 22.

ABC (1990a), "Sixteen well trained Soweto builders receive certificates", January/February, pp. 12-13.

ABC (1990b), "The second AGM of the Professional Builders Federation - chairman calls for more market information", reprint of speech given at the second AGM (Theme for the day - Black Economic Empowerment) of the PBA at Johannesburg, 18 November 1989, by president Linda Nyembe, January/February, pp. 7-10.

ABC (1990c), "Training, finance and advice now available", October/November, pp. 29-30.

ABC (1990d), "Strong links forged between black and white builders' associations", October/November, p. 11.

ABC (1990e), "Palm Springs - an SAHT project", March April, pp. 20-21.

ABC (1990f), "Development programme scored in SAHT Small Builder Contest", October/November, p. 18.

ABC (1990g), "Housing Trust helps to train small builders", October/November, p. 25.

ABC (1991a), "Now the professional small builder can compete thanks to the new CDA scheme", April/May, p. 12.

ABC (1991b), "Very positive TABA AGM", April/May, p. 7.

ABC (1991c), "Winners of the S.A. Housing Trust's small builder of the year competition", December/January, pp. 4-6.

ABC (1991d), "A profile of the S.A. Housing Trust", September, pp. 24-25.

ABC (1991e),   "Reef training centre for builders could be for you", December/January, p. 12.

African National Congress (ANC) (1993),   *Ready to Govern*. African National Congress policy guidelines for a democratic South Africa adopted at the National Conference 28-31 May 1992.

Alexander, G.D. (1990),   "A partnership that 'over-extended' itself but was rescued from failure", *African Building Contractor*, January/February, pp. 22-30.

Archer, S. (1988),   "'The best of all possible worlds'. debating alternative economic systems in South Africa", *Sash*, December, pp. 5-12.

Bloemfontein Building Industry Development Foundation (BBIDF) (1989),   "Bouaannemersontwikkelingsprogram BAO (Vlak 1) -Kursus vir klein bouaannemers - Verslag 1989", report obtained from Professor B. Verster, University of the Orange Free State.

BBIDF (1990),   "Bouaannemersontwikkelingsprogram - Kursus vir klein bouaannemers - Verslag 1990", report obtained from Professor B. Verster, University of the Orange Free State.

BBIDF (1991),   "Bouaannemersontwikkelingsprogram BOA (Vlak 1) -Kursus vir klein bouaannemers - Verslag 1991", report obtained from Professor B. Verster, University of the Orange Free State.

Building Industries Federation of South Africa (BIFSA) (1983),   BIFSA Annual Report.

BIFSA (1984),   BIFSA Annual Report.

BIFSA (1985),   BIFSA Annual Report.

BIFSA (1986),   BIFSA Annual Report.

BIFSA (1987),   BIFSA Annual Report.

BIFSA (1988a),   BIFSA Annual Report.

BIFSA (1988b),   Circular 1988/11 - ADM. 201.01. Correspondence to members detailing revised by-laws referenced RMA/11b, 20 January 1988.

BIFSA (1989),   BIFSA Statistical Yearbook.

BIFSA (1990),   BIFSA Statistical Yearbook.

BIFSA (1992),   BIFSA Statistical Yearbook.

BIFSA,   *BIFSA Official Handbook* (loose leaved - no identifiable publication date).

Brus, W. and Laski, K. (1989),   *From Marx to the Market : Socialism in search of an economic system*. Clarendon : Oxford.

Bukharin, N.I., Debonorin, A.M., Uranovsky, Y.M., Vavilov, S.I., Komarov, V.L. and Tiumeniev, A.I. (1936),   *Marxism and Modern Thought*. George Routledge & Sons : London.

Burgess, R. (1978),   "Petty commodity housing or dweller control? A critique of John Turner's views on housing policy", *World Development*, Vol.6, No. 9, pp. 1105-1133.

Calcopietro, C.M. (1992),   "Institutions supporting small-scale builders in South Africa"; unpublished paper obtained from the author, pp. 1-42.

Census of Construction (1985),   See Republic of South Africa, (1985).

Courier (1991),   "Decade of Achievement", Vol. 9, No. 1, p. 3.

Creighton, P. (1992),   "The Hodeco Operation", unpublished address given to M.B.S. (Pty) Ltd. at Downtown Inn on 25 January 1992, transcript obtained from the author.

Currie, L. (1971),   "The exchange constraint on development : A partial solution to the problem", *Economic Journal*, 81, pp. 886-903

Currie, L. (1983),   "Housing as an Instrument of Macro-economic Policy", *Habitat International*, Vol. 7, No. 5/6, pp. 165-171.

Cohen, S. (1974),   *Bukharin and the Bolshevik Revolution. A Political Biography 1888-1938*. Wildwood House : London.

Cohen, S. (1986),   *Rethinking the Soviet Experience*. Oxford University Press : New York.

Community and Urban Services Support Project (CUSSP) (1993),   Information brochure obtained from Cape Town office of CUSSP.

Duncan, K. (1991),   "An integrated approach to the economic empowerment of small builders in South Africa", article obtained from the author (excerpts from an article written in August 1991 proposing the repositioning in the market of the SA Housing Trust's Small Builders Department).

Duncan, K. (1992a),   "Economic empowerment through the SA Housing Trust's new Small Builder Programme", - article obtained from the author (an article written for the Sowetan in January 1992).

Duncan, K. (1992b),   "Small Builders Department", article obtained from the author (article written for SA Housing Trust's in-house magazine in February 1992).

Edmonds, G.A. and Miles, D.W.J. (1984),   *Foundations For Change. Aspects of the construction industry in developing countries*. Intermediate Technology Publications : London.

Edwards, C. (1985),   *The Fragmented World. Competing perspectives on trade, money and crisis*. Methuen : New York.

Engels, F. (1979),   *The Housing Question*. Progress Publishers : Moscow.

Entrepreneurial Development (Southern Africa) (EDSA) (1990),   Annual Report 1990.

EDSA (1991),   Annual Report 1991.

EDSA (1992),   Annual Report 1992.

Ferreira, F.H. and Potgieter, J.F. (1989),   "The venture capital industry in south africa and the potential for participation by black entrepreneurs", Institute of Social and Economic Research, Development Studies Unit, Rhodes University, Grahamstown, Working Paper No. 48, pp. 53-55.

Financial Mail (1986),   "Building cartel : coming to terms", Vol. 102, No. 11, 12 December, p. 47.

Fraser, N. (1989),   "Under-quoting is a formula for disaster", *African Building Contractor*, September/October, pp. 12-16.

Griffin, C. (1990),   "Don't get caught-out by price increases", *African Building Contactor*, January/February, pp. 18-19.

Habitat International Coalition (HIC) (1988),   *Building Community. A third world case book*. Turner, J.F.C. Ed., Building Community Books : London.

Harrington, M. (1977),   *The Twilight of Capitalism*. Macmillan : London.

Hendler, P. (1989),   *Politics on the Home Front*. South African Institute of Race Relations : Johannesburg.

Hendler, P. (1993),   "Analysis, critique and strategic implications of the De Loor Report", report prepared for Planact.

Herbert, J.D. (1979),   *Urban Development in the Third World. Policy guidelines*. Praeger Publishers : New York.

Housing in Southern Africa (HSA) (1989a),   "Outgoing chairman sees '89 as a year filled with many challenges and opportunities", April, pp. 9-10.

HSA (1989b),   "South Africans spend billions on additions and extensions", May, p. 12.

HSA (1989c),   "Builders urged to form closer ties with local authorities", August, p. 18.

HSA (1989d),   "Investing in the future of the country", November/December, pp. 20-23.

HSA (1989e),   "What is the SA Housing Trust?", January, pp. 13-15.

HSA (1990a),   "Whites not keen to promote Blacks", February, p. 33.

HSA (1990b),   The ANC and COSATU on the provision of new affordable homes in a post-Apartheid R.S.A", August, pp. 4-5.

HSA (1991a),   "The time for theorising and politicising is over - Hardy", February, pp. 11-13.

HSA (1991b),   "The low-cost market is going to become the dominant market - SAHT", February, p. 20.

HSA (1991c),   "The era of the large building contractor is over - survey", March, p. 9.

HSA (1991d),   "Housing for all - no longer just a dream", November/December, pp. 18-19.

HSA (1991e),   "The DBSA plays a vital role in securing a better future for all", April, p. 21.

International Labour Organisation   *Guide-Lines for the Development of Small-Scale Construction Enterprises.* International
(ILO) (1987),   Labour Organisation : Geneva.

Kaser, M. (1970),   *Soviet Economics.* World University Library. McGraw-Hill : New York.

Kay, G. (1975),   *Development and Underdevelopment. A Marxist Analysis.* Macmillan : London.

Koenigsberger, O. (1986),   Third World Housing Policies since the 1950s", *Habitat International*, Vol. 10, No. 3, pp. 27-32.

Krafchik, W. (1990),   *Small-Scale Enterprises, Inward Industrialisation, and Housing : A Case Study of Subcontractors in the Cape Peninsula Low-Cost Housing Industry.* Unpublished Masters thesis, School of Economics, University of Cape Town.

Krafchik, W. (1991),   "Growth possibilities and constraints in the Western Cape construction industry", in *Working Papers on the Economy of the Western Cape*, Paper 2, pp. 1-24.

Kwazulu Training Trust (KTT)   Annual Report 1992.
(1992),

Kwazulu Finance and Investment   Annual Report 1992.
Corporation Limited (KFC) (1992),

Lekachman, R. and Van Loon, B.   *Capitalism for Beginners.* Writers and Readers Publishing Co-op. : London.
(1981),

Lipton, M. (1977),   *Why Poor People Stay Poor. A study of urban bias in world development.* Temple Smith : London.

Mabe, S. (1989),   Cited in *African Building Contractor* (1989c), "NAFCOC and the Professional Builders Federation hold builders workshop", reprint of speech given at seminar (The Black Building Industry - Prospects for economic upliftment) by Linda Nyembe, PBA chairman, at Pretoria 27 June 1989, September/October, p. 17.

Maitra, P. (1986),   *Population, Technology and Development. A critical analysis.* Gower : Aldershot.

Mandel, E. (1989),   *Beyond Perestroika. The future of Gorbachev's USSR.* Verso : London.

Marx, K. (1973),   *Grundrisse: Foundations of the Critique of Political Economy.* Penguin : Harmondworth; cited by Nove (1986a), *Marxism and "Really Existing Socialism".* Harwood Academic Publishers : London.

Marx, K. (1899),   *Value, Price and Profit : Addressed to working men.* George Allen & Unwin : London.

McConnell, J.W. (1943),   *The Basic Teachings of the Great Economists.* The New Home Library : New York.

Merrifield, A. (1992),   "Private sector involvement in South Africa's low-income housing market since the late 1980s". Unpublished report, Department of Property Development and Construction Economics, University of Natal.

Mngomezulu, S. (1988),   "Builders loans", *African Building Contractor*, September/October, p. 10.

Moll, P. (1990),   *The Great Economic Debate.* Skotaville Publishers : Johannesburg.

Moser, C.O.N. (1978),   Informal sector or petty commodity production : Dualism or dependence in urban development", *World Development*, Vol.6, No. 9, pp. 1041-1064.

Motlanthe, P. (1990),   "Critical needs of the emerging contractor", *African Building Contractor*, January/February, pp. 15-16.

National African Federated
Chamber for the Building Industry
(NAFBI) (1991),

Constitution - obtained from Mr. C. Petersen of NAFBI, pp. 1-13.

NAFBI (1992),

"Proposal - Establishment of Technical Aid Centres", unpublished circular - obtained from Mr. C. Petersen of NAFBI, pp. 1-3.

National Building Research Institute
(NBRI) (1987),

Low-Cost Housing. CSIR : Pretoria.

Nientied, P and van der Linden, J.
(1985),

"Approaches to low-income housing in the third world: some comments", *International Journal of Urban and Regional Research*, 9 (3), pp. 311-328.

Nonyane, D. (1988),

"A new house needed every 40 seconds!", *African Building Contractor*, September/October, p. 2.

Nove, A. (1983),

The Economics of Feasible Socialism. George Allen & Unwin : London.

Nove, A. (1986a),

Marxism and "Really Existing Socialism". Harwood Academic Publishers : London.

Nove, A. (1986b),

Socialism, Economics and Development. Allen & Unwin : London.

Nove, A. (1989),

Glasnost' in Action. Cultural Renaissance in Russia. Unwin Hyman : Boston.

Padi, D. (1988a),

"Success stories", *African Building Contractor*, September/October, pp. 3-9.

Padi, D. (1988b),

"Success stories", *African Building Contractor*, November/December, pp. 2-3.

Padi, D. (1988c),

"A professional A.G.M. for a truly professional organisation", *African Building Contractor*, November/December pp. 6-7.

Padi, D. (1989a),

"Success stories", *African Building Contractor*, May/June, pp. 2-5.

Padi, D. (1989b),

"Success stories", *African Building Contractor*, August, pp. 3-6.

Padi, D. (1989c),

"Success stories", *African Building Contractor*, September/October, pp. 2-8.

Padi, D. (1990a),

"Success stories", *African Building Contractor*, January/February, p. 3.

Padi, D. (1990b),

"Success stories", *African Building Contractor*, March/April, p. 2-8.

Padi, D. (1990c),

"Success stories", *African Building Contractor*, May/June, p. 2-8.

Preobrazhensky, E.A. (1980),

The Crisis of Soviet Industrialization. Macmillan : London.

Republic of South Africa (1985),

"Census of Construction, 1985", Report No. 50-01-01 (1985).

Republic of South Africa (1987a),

Regulation Gazette No. 4114, Government Gazette, Vol. 265, No.10844, 31 July 1987, pp. 1-35.

Republic of South Africa (1987b),

Regulation Gazette No. 4078, Government Gazette, Vol. 262, No.10720, 24 April 1987, pp. 1-17.

Republic of South Africa (1989),

Regulation Gazette No. 4332, Government Gazette, Vol. 285, No.11766, 23 March 1989, pp. 1-48.

Republic of South Africa (1992a),

Regulation Gazette No. 4838, Government Gazette, Vol. 321, No. 113852, 20 March 1992, pp. 9-28.

Republic of South Africa (1992b),

"SA Statistics, 1992". Central Statistical Services, Pretoria.

Riddell, R.C. (1987),

Foreign Aid Reconsidered. James Currey : London.

Rolfe, S.M. (1990),

The Stimulation of Small-Scale Enterprises in South Africa Through the Process of Subcontracting. Unpublished Honours thesis, School of Economics, University of Cape Town.

Roukens De Lange, A. (1990),

"The Impact of Provision of Low-Cost Housing on the South African Economy", report published by the Institute for Futures Research, University of Stellenbosch, pp. 11-17.

SA Statistics (1992),

See Republic of South Africa, (1992b).

Singer, P. (1980),

Marx. Oxford University Press : Oxford.

Slingsby, M. (1986),   "Community Development Support Programmes for Housing Projects - A Problem-Solving Approach", *Habitat International*, Vol. 10, No. 3, pp. 65-71.

South African Housing Advisory Council (SAHAC) (1992),   "Housing in South Africa : Proposals on a policy and strategy" (a.k.a. The De Loor Report).

Strassmann, P.W. (1978),   *Housing and Building Technology in Developing Countries*. Michigan State University International Business and Economic Studies : Michigan.

Suter, R.W. (1993),   *The Role of Architects in Developing Countries : A Focus on Rural Areas*. Unpublished undergraduate term paper, School of Architecture and Planning, University of Cape Town.

Taylor, J.G. (1979),   *From Modernization to Modes of Production. A critique of the sociologies of development and underdevelopment*. Macmillan : London.

Tholiwe Homes (TH) (1992),   Internal reports, *etc.* obtained from Tholiwe Homes.

Turner, J.F.C. (1972),   "Housing as a verb", in *Freedom To Build*, J.F.C. Turner and R. Fichter (eds.). Macmillan : New York.

Turner, J.F.C. (1978),   "Housing in three dimensions : Terms of reference for the housing question redefined", *World Development*, Vol.6, No. 9, pp. 1135-1145.

United Nations (1985),   *Relationship Between Housing and the National Economy*. Synthesis report on the seminar held in Prague (Czechoslovakia) 10-14 May 1982. United Nations : New York.

University of Cape Town Careers Office (1991),   *Bridging The Gap 1992 - A Guide to Practical & Educational Courses in the Western Cape*. UCT Careers Office : Cape Town.

Urban Foundation (UF) (1988),   "The Urban Foundation Annual Review 1988", pp. 1-29.

UF (1989a),   "Role and Achievements", unpublished internal pamphlet, pp. 1-14.

UF (1989b),   "The Urban Foundation Annual Review 1989", pp. 1-21.

UF (1990),   "The Urban Foundation Annual Review 1990", pp. 1-39.

UF (1991),   "The Urban Foundation Annual Review 1991", pp. 1-36.

UF (1992),   "The Urban Foundation in 1992", unpublished internal pamphlet, pp. 1-9.

Van Staden, T. (1993),   "A holistic approach to SME development through grassroots community participation", *Proceedings of the 21st IAHS World Housing Congress*, Vol. 1, pp. 579-590.

Van Niekerk, C.J. du P and Hutton, R. (1993),   "Affordable 100% steel housing : A manufacturer's experience in the RSA", *Proceedings of the 21st IAHS World Housing Congress*, Vol. 1, pp. 567-568.

Watermeyer, R.B. (1993),   "Community-based construction : Mobilising communities to construct their own infrastructure", *Proceedings of the 21st IAHS World Housing Congress*, Vol. 1, pp. 603-614.

Wakely, P. (1986),   "The Devolution of Housing Production : Support and Management", *Habitat International*, Vol. 10, No. 3, pp. 53-63.

Wells, J. (1972),   "The construction industry in East Africa", Paper 72.2, Economic Research Bureau, University of Dar Es Salaam.

Wells, J. (1986),   *The Construction Industry in Developing Countries : Alternative Strategies for Development*. Croom Helm : Beckenham

Wilber, C.K. (1969),   *The Soviet Model and Underdeveloped Countries*. University of North Carolina Press : Chapel Hill.

World Bank (1984),   *The Construction Industry - Issues and strategies in developing countries*. The World Bank : Washington.

APPENDIXES

Appendixes A to G are presented in Volume 1

Appendixes H to Q are presented in Volume 2

DETAILED INDEX OF APPENDIXES

NOTES ON TECHNICAL SCHOOLING AND TERTIARY EDUCATION
IN SOUTH AFRICA

The contents of this section describe technical and tertiary education available in South Africa. The purpose of its inclusion is to facilitate the reader's interpretation of statements made by interviewees regarding their education, training or qualifications.

The institutions which offer post-school qualifications in South Africa are : Technical colleges, Technikons, Teachers Training Colleges and Universities. Only those which are relevant to the construction industry are described in detail.

(i)   TECHNICAL COLLEGES

The function of technical colleges. There are approximately 120 technical colleges in South Africa, each tailoring its curriculum to suit the training needs of its immediate community. The fundamental function of a technical college is to provide career training for school-leavers at both the pre-tertiary as well as the tertiary levels of the South African educational system. Programmes focus on training for only those skills which will be required for the chosen career and usually exclude subjects of a peripheral nature. Training programmes provide both the theoretical and practical knowledge required in vocations and are structured as sandwich courses, requiring alternate spells of college attendance and acquisition of experience in the work place.

Types of courses offered at technical colleges. Both non-formal as well as formal education courses are presented.

Non-formal education programmes are offered, usually as short courses, and syllabuses are designed to suit the specific training needs of the local community, industry and commerce.

The formal courses have approved (controlled) curricula which lead to the acquisition of either a national Certificate or a National N-Diploma. Training is offered on either a full-time or a part-time basis in the following broad fields of study : Technical/Engineering, Hair Care, Commercial, Educare, Community Services and the Fine Arts. Only the technical courses relevant to building construction are described below.

Training of artisans. The major function of the technical colleges is the training of tradesmen or artisans and courses are offered which prepare students for careers as bricklayers, plasterers, plumbers, carpenters, joiners, painters, decorators and installation electricians.

Levels of study. These trade-oriented courses are offered at seven levels, namely N1 to N6. Each level of study generally consists of four subjects and is presented over a ten to eleven week period. A national examination is then written at the college during the twelfth and thirteenth weeks of the trimester. Students are issued with a certificate for each level successfully completed. Levels N1, N2 and N3 are pre-tertiary, or technical schooling levels and are the approximate equivalents of standards 8, 9 and 10 (senior certificate or matric) of normal secondary schooling. A student who has passed the N3 level and also passes the two official languages (English and Afrikaans) is awarded a senior certificate. Levels N4, N5 and N6 are tertiary levels of study and once twelve subjects have been passed at these levels, a National N-Diploma is awarded.

Entrance requirements. The minimum entrance requirement for the N1 level of study is a standard 7 school leaving certificate preferably with mathematics and science/physics. Colleges offer a ten week bridging course for students who do not qualify to enter at N1 level. This course includes mathematics, science, technical drawing and general trade theory.

Trade tests. Once the N2 level of study has been passed, students qualify to register for a trade test. Once this test is

---

[1]   Source : University of Cape Town Careers Office, (1991).

passed, artisan status is awarded. Students in the electrical field gain entry to the qualifying examination for the Government Certificate of Competency after passing eighteen subjects at tertiary level.

(ii) TECHNIKONS

The function of technikons. The primary task of technikons is to provide education and training for students who w to enter the labour market with specific vocational skills and a knowledge of technology. Emphasis is placed on th application of theory and a strong vocational bias aims to ensure that successful students are immediately useful ar capable of being productive when they enter the job market.

Types of courses offered at technikons. A wide range of courses are offered at the various colleges covering almo every conceivable area of study. Those which are relevant to the construction industry are : Architecture/Architec technology, Building Surveying, Building technology, Civil Engineering, Electrical Engineering and Housing Development and Management. These courses, as a part of the curriculum, require that students obtain practical experience, or in-service training. Depending on the course, three to eighteen months may have to be spent in a w situation.

The following types of qualifications can be obtained from technikons :-

National Certificate (one year)
National Higher Certificate (two years)
National Diploma (3 years)
National Diploma (4 years)
National Post-Diploma Certificate (one year's study after obtaining National Diploma)
National Higher Diploma (one year's study after obtaining National Diploma)
Masters Diploma in Technology (two year's study after obtaining National Diploma)
Laureatus in Technology (two year's study after obtaining Masters Diploma in Technology)

Entrance requirements. The minimum requirement for entrance to a technikon is a senior certificate (matric or standard 10) with six subjects. In addition, specific matric subjects may be required, for example, mathematics, science, *etc.*. Many technikon applicants reportedly hold matriculation exemption certificates (the minimum requirement for entrance to a university) and aggregates of 60% or higher are common. In effect this places acceptance at certain technikons beyond the reach of many applicant who would qualify to enter.

(iii) UNIVERSITIES

Degree courses in Building Management and Quantity Surveying are offered at six South African Universities. Th take the form of either four or five year programmes. Minimum entrance requirements usually include a matricula exemption certificate and a minimum of 50% in mathematics and a science subject.

APPENDIX B

NOTES ON THE INDUSTRIAL COUNCIL SYSTEM AND AGREEMENTS

B.1 The Labour Relations Act (1956) and Industrial Councils[2]

The establishment and registration of Industrial Councils is provided for in the Labour Relations Act (1956) (LRA). Industrial councils are standing bodies made up of an equal number of employer and employee representatives. Groupings of employers and employees who are party to the Council meet in this forum to negotiate and formulate agreements on matters of common interest. Parties negotiate and draft their own constitution for the proposed Industrial Council and submitted this to the Registrar to fulfil the registration requirements.

The principal function of the Industrial Councils is to maintain stable labour relations through the formulation of methods for the settlement of disputes. Procedures for resolving disputes between employer and employees members of the Council are drafted into an Industrial Council Agreement and usually specify a number of meetings between the parties concerned, with the option of arbitration. No legal strike or lock-out may take place in an industry or area where the Industrial Council has jurisdiction until these procedures have been followed.

Section 24(i) of the LRA lists the following matters of common interest that may be addressed in an agreement:

- minimum wages payable to defined categories of employees
- regulations governing overtime
- closed shop provisions
- the establishment of holiday, medical, death and training funds for employees

The agreement is automatically binding on the parties to the Council. However, the Minister of Manpower may extend the scope of the agreement to apply to all employers and employees in a defined geographical area and industry, regardless of whether or not they are party to the Council. Section 53 of the LRA defines that any individual who contravenes the provisions of the agreement is guilty of a criminal offence. It is the duty of the Industrial Council to ensure that all parties over whom it has jurisdiction comply with the provisions of the agreement. If deemed necessary, the Minister of Manpower may appoint an official to police the agreement.

B.2 Industrial Council Agreements[3]

Regionally established Industrial Councils administer the various agreements formulated by groups of employer and employee bodies within that region. Thus, one Industrial Council could possibly administer any number of agreements, but each would apply to a defined geographical area in which the parties to the agreements usually operate. These agreements are binding for a specific period, whereafter the agreement is published anew with any amendments.

The subjects of the survey for this study fall under the jurisdiction of the Industrial Councils for the Building Industry, Port Natal and the Building and Monumental Masonry Industries (Transvaal), as the case may be. The Port Natal council controls one agreement for the Durban area, while the Transvaal council administers two agreements - one for unskilled employees and one for all other employees. Only those aspects of the agreement which are pertinent to this report are highlighted and the agreements themselves should be consulted if further detail is required.

---

[2]   Source: Krafchik (1990: 218-223)

[3]   Republic of South Africa (1987a, 1987b and 1992a)

(a) *Who the agreements affect:* As noted above, the Minister may decree that the agreement be adhered to by individuals/firms other than those who were party to the agreement. Both the Port Natal and the Transvaal agreements have been so extended. This is seen in section (b) of the introduction to each agreement :- "in terms of [the Labour Relations Act, I the deputy Minister of Manpower] declare that the provisions of the said agreement [excluding certain clauses] shall be binding...upon all employers and employees, other than those [who are members of the employers' organisations and trade unions] who are engaged or employed in [the building industry in the areas specified]." It is this provision that makes it a statutory offence for any employer or employee to fail to comply with the agreement, albeit that they were not party to the agreement.

(b) *Registration and deposit/guarantee:* To operate legally in the building industry a firm must register with the Industrial Council with the Council. It is a statutory offence for a firm to be unregistered while operating in the industry. A requirement of registration is that the employer must lodge a deposit with the Industrial Council to cover liabilities in the event of an insolvency.

In the case of the main (that is, other than unskilled) Transvaal agreement, the amount of this deposit must be equal to the total of 2 weeks' wages (statutory minimum wages) for each employee as well as two weeks' employer contributions as prescribed by the agreement. This deposit shall not be less than R 500 per employee, in the case of weekly and bi-weekly paid employees and not less than R 5 000 per employee, in the case of monthly paid employees.

In the case of the Transvaal agreement for unskilled employees, the amount of this deposit in respect of weekly paid employees must be equal to the total of 2 weeks' wages (statutory minimum wages) for each employee as well as two weeks' employer contributions as prescribed by the agreement. This deposit shall not be less than R 1 000 per employee. In the case of fortnightly paid workers the amount is for 3 weeks' wages and contributions, subject to a minimum of R 1 000, while for monthly paid employees, the amount is for 5 weeks' wages and contributions, subject to a minimum of R 2 000.

The Port Natal agreement provides for the employer to pay a wage guarantee in respect of each employee, equal to 2 weeks' wages and inclement weather allowance and 2 weeks' stamps, subject to a minimum of R 750.

(c) *Minimum wages:* Basic minimum wages are specified in the agreements for each class of employee - with the exception being that in the Port Natal agreement, wages of 'general workers' are not prescribed. These minimum wages are revised periodically during the tenure of each agreement. The following are the current (1992) minimum basic wages for general workers and artisans:

	MAIN TRANSVAAL AGREEMENT	TRANSVAAL AGREEMENT FOR UNSKILLED EMPLOYEES	PORT NATAL AGREEMENT
GENERAL WORKERS	N/A	R 659	R 608[*]
ARTISANS	R 1 818	N/A	R 1 941

All wages are reflected as **monthly wages**

(d) *Industrial Council Stamps:* Over and above the minimum basic wage employers are liable for weekly contributions in respect of holiday, pension, sick and death benefit funds, *etc.* for employees and for the operating expenses of the Industrial Council and the Building Industry Training Fund. As with minimum wage amounts, the amounts of these contributions are revised periodically. Payments are made to the Industrial Council, which issues stamps in respect thereof to employers. The employer passes the stamps on to his employees who place them in a contribution booklet (which booklet remains the property of the Industrial Council). The booklets are collected annually by the Industrial Council and holiday pay and payments from other funds are determined.

---

[*] Although not prescribed, G. Norvell of the University of Natal reported that R 3.51 per hour, or R 608.40 per month, can be assumed a reliable average.

(e) Prohibited employment: Both of the Transvaal agreements and the Port Natal agreement prohibit the use of unskilled labour for performing skilled work. Only **registered** tradesmen, artisans, craftsmen, master craftsmen, foremen or apprentices are permitted to perform skilled work. Interestingly, the agreements state that even if unskilled employees are used for skilled work they shall be paid the appropriate wages as if they had been skilled. Many would argue that the reservation of skilled work for registered skilled employees contradicts the spirit of the Port Natal agreement, which defines its purpose as:

> "to recognise the level of skill of every employee, to provide opportunities for his further progress and to establish minimum levels of remuneration and other conditions of employment for employees in the building industry without in any way restricting entrepreneurial initiative and employment opportunities."

'Skilled work' is defined in the main Transvaal agreement as:

> "'skilled work' means any work in the Building and Monumental industries which may be performed by an employee as defined under 'skilled employee'"

The definition given for 'skilled employee' is:

> "'skilled employee' means any general foreman, foreman, master craftsman, craftsman, artisan, craftman's assistant, apprentice, specified skills employee, learner specified skills employee, trainee master craftsman, trainee craftsman's assistant and an employee engaged in an ancillary trade."

The definitions of these terms given in the Transvaal agreement for unskilled employees are very similar.

In the Port Natal agreement, 'skilled work' is defined as:

> "... work of a skilled nature which is normally and customarily performed by a person who has served under a contract of apprenticeship or completed a period of training in terms of the Manpower Training Act, 1981, in any of the trades designated in the Act..."

SUPPORTING TABLES AND FIGURES FOR CHAPTER 3

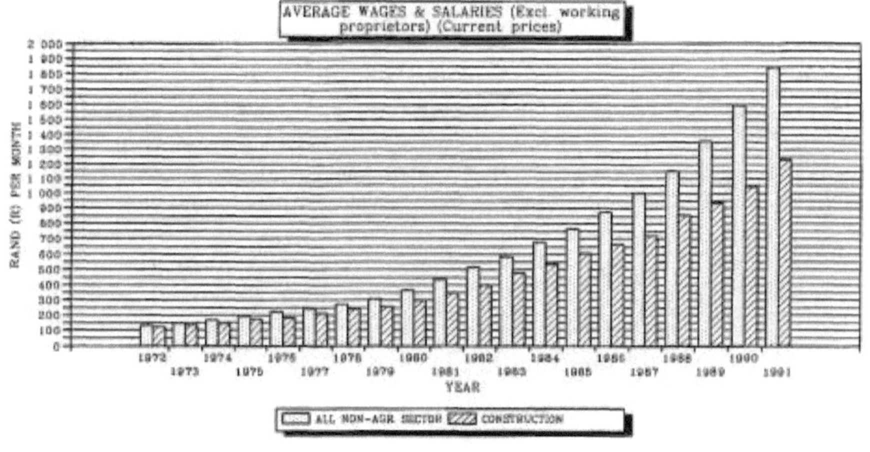

Source : SA Statistics (1992)

Figure C1: Average Construction and Average Total Non-Agricultural Wages and Salaries

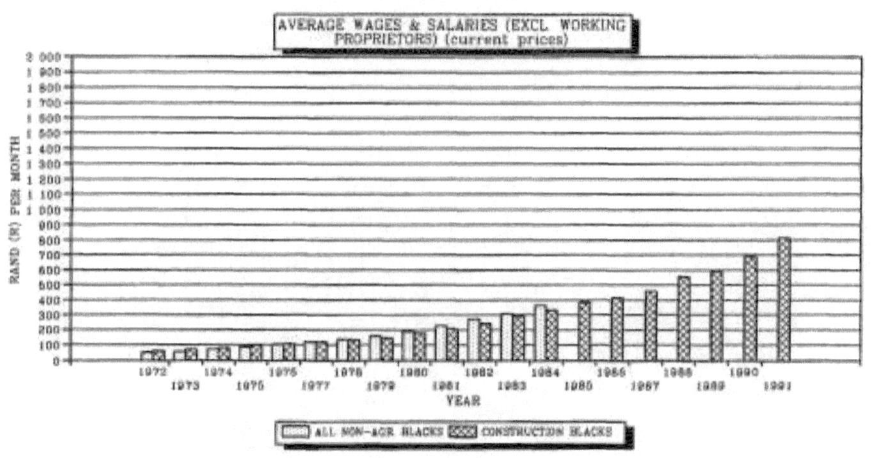

Source : SA Statistics (1992)

Figure C2: Average Wages and Salaries of Construction Blacks and Total Non-Agricultural Blacks

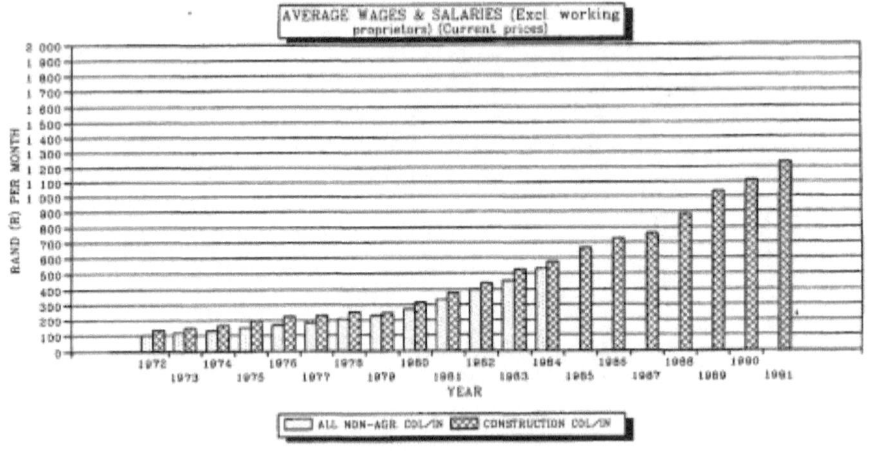

Source : SA Statistics (1992)

Figure C3: Average Construction and Total Non-Agricultural Coloured & Indian Wages and Salaries

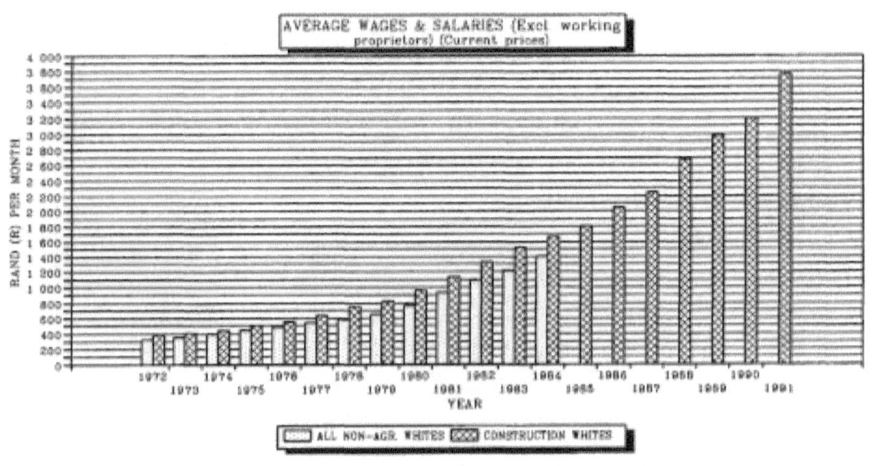

Source : SA Statistics (1992)

Figure C4: Average Construction and Total Non-Agricultural White Wages and Salaries

TABLE C1 – CONSTRUCTION AND GDP

YEAR	GDP	CONSTR.	CONSTR./ GDP	GROWTH ON PREVIOUS YEAR	
	Million (R)			GDP	CONSTR
1960	44 238	1 343	3.0%	–	–
1961	45 987	1 327	2.9%	4.0%	-1.2%
1962	48 807	1 355	2.8%	6.1%	2.1%
1963	52 067	1 528	2.9%	6.7%	12.8%
1964	55 813	1 923	3.4%	7.2%	25.9%
1965	59 038	2 282	3.9%	5.8%	18.7%
1966	61 654	2 259	3.7%	4.4%	-1.0%
1967	65 277	2 462	3.8%	5.9%	9.0%
1968	68 331	2 629	3.8%	4.7%	6.8%
1969	72 168	3 004	4.2%	5.6%	14.3%
1970	75 891	3 446	4.5%	5.2%	14.7%
1971	79 203	3 910	4.9%	4.4%	13.5%
1972	80 735	4 254	5.3%	1.9%	8.8%
1973	83 814	4 262	5.1%	3.8%	0.2%
1974	86 319	4 726	5.4%	5.4%	10.9%
1975	90 091	4 841	5.4%	2.0%	2.4%
1976	92 675	4 654	5.0%	2.9%	-3.9%
1977	92 657	4 403	4.8%	-0.0%	-5.4%
1978	95 298	4 041	4.2%	2.9%	-8.2%
1979	99 032	4 095	4.1%	3.9%	1.3%
1980	105 121	4 441	4.2%	6.1%	8.4%
1981	110 387	4 687	4.2%	5.0%	5.5%
1982	109 634	4 502	4.1%	-0.7%	-3.9%
1983	107 244	4 256	4.0%	-2.2%	-5.4%
1984	112 859	4 363	3.9%	5.2%	2.5%
1985	112 448	4 144	3.7%	-0.4%	-5.0%
1986	112 458	3 880	3.5%	0.0%	-6.4%
1987	114 220	3 557	3.1%	1.6%	-8.3%
1988	118 556	3 672	3.1%	3.8%	3.2%
1989	121 216	3 963	3.3%	2.2%	7.9%
1990	120 488	3 956	3.3%	-0.6%	-0.2%
1991	119 596	3 777	3.2%	-0.7%	-4.5%

GDP   – Gross Domestic Product
CONSTR.   – Construction's contribution to GDP
CONSTR/GDP   – Construction divided by GDP
R   – South African Rands

Factor incomes at constant (1985) prices
Source : SA Statistics (1992: 21.6)

TABLE C2 - NON-AGRICULTURAL EMPLOYMENT

Excluding working proprietors					Including working proprietors		
YEAR	NON-AGRICULTURAL EMPLOYMENT			GROWTH ON PREVIOUS YEAR		WORKING PROPRIETORS	
	ALL SECT	CONSTR.	CONSTR. %	ALL SECT	CONSTR.	TOTAL	WKNG PR	% OF TOT
1972	3 936 550	343 958	8.7%			349 393	5 435	1.6%
1973	4 197 036	403 700	9.6%	6.6%	17.4%	-	-	-
1974	4 404 060	476 178	10.8%	4.9%	18.0%	483 003	6 825	1.4%
1975	4 510 930	483 600	10.7%	2.4%	1.6%	-	-	-
1976	4 701 871	466 409	9.9%	4.2%	-3.6%	474 249	7 840	1.7%
1977	4 595 411	376 700	8.2%	-2.3%	-19.2%	-	-	-
1978	4 505 241	315 857	7.0%	-2.0%	-16.2%	322 403	6 546	2.0%
1979	4 584 933	315 100	6.9%	1.8%	-0.2%	-	-	-
1980	4 787 343	364 164	7.6%	4.4%	15.6%	370 658	6 494	1.8%
1981	5 006 151	414 100	8.3%	4.6%	13.7%	-	-	-
1982	5 113 263	446 866	8.7%	2.1%	7.9%	456 202	9 336	2.0%
1983	5 043 375	428 100	8.5%	-1.4%	-4.2%	-	-	-
1984	5 109 660	424 300	8.3%	1.3%	-0.9%	-	-	-
1985	5 063 393	410 100	8.1%	-0.9%	-3.3%	420 966	10 866	2.6%
1986	5 066 462	402 500	7.9%	0.1%	-1.9%	-	-	-
1987	5 143 005	407 600	7.9%	1.5%	1.3%	-	-	-
1988	5 220 544	413 800	7.9%	1.5%	1.5%	-	-	-
1989	5 261 562	417 200	7.9%	0.8%	0.8%	-	-	-
1990	5 238 329	417 500	8.0%	-0.4%	0.1%	-	-	-
1991	5 139 149	391 000	7.6%	-1.9%	-6.3%	-	-	-

ALL SECT   - Total non agricultural employment including construction
CONSTR   - Construction employment excluding working proprietors
CONSTR %   - Construction proportion of total
TOTAL   - Construction employment including working proprietors
WKNG PROP - Construction working proprietors
% OF TOT   - Proportion construction working proprietors of total construction

Source : SA Statistics (1992: 7.8-7.13 & 13.4) Census of Construction (1985:1)

TABLE C3 - CONSTRUCTION EMPLOYMENT BY RACE GROUP

Excluding working proprietors					GROWTH ON PREVIOUS YEAR			
YEAR	TOTAL	BLACK	COL/IND	WHITE	TOTAL	BLACK	COL/IND	WHITE
1972	343 958	235 111	56 645	52 202	-	-	-	-
1973	403 700	261 700	64 200	57 800	17.4%	19.8%	13.3%	10.7%
1974	476 178	334 477	75 320	66 381	18.0%	16.7%	17.3%	14.8%
1975	483 600	342 100	75 800	65 700	1.6%	2.3%	0.6%	-1.0%
1976	466 409	330 359	74 254	61 796	-3.6%	-3.4%	-2.0%	-5.9%
1977	376 700	261 500	63 000	52 200	-19.2%	-20.8%	-15.2%	-15.5%
1978	315 857	217 694	53 702	44 461	-16.2%	-16.8%	-14.8%	-14.8%
1979	315 100	215 600	55 400	44 100	-0.2%	-1.0%	3.2%	-0.8%
1980	364 164	255 814	63 370	44 980	15.6%	18.7%	14.4%	2.0%
1981	414 100	294 600	70 000	49 500	13.7%	15.2%	10.5%	10.0%
1982	446 871	318 828	76 446	51 592	7.9%	8.2%	9.2%	4.2%
1983	428 100	304 300	72 900	50 900	-4.2%	-4.6%	-4.6%	-1.3%
1984	424 300	299 900	73 300	51 100	-0.9%	-1.4%	0.5%	0.4%
1985	410 100	290 758	69 608	49 734	-3.3%	-3.0%	-5.0%	-2.7%
1986	402 500	286 100	67 700	48 700	-1.9%	-1.6%	-2.7%	-2.1%
1987	407 600	293 100	66 600	47 900	1.3%	2.4%	-1.6%	-1.6%
1988	413 800	298 600	66 800	48 400	1.5%	1.9%	0.3%	1.0%
1989	417 200	302 200	66 900	48 100	0.8%	1.2%	0.1%	-0.6%
1990	417 500	303 000	66 200	48 300	0.1%	0.3%	-1.0%	0.4%
1991	391 000	280 800	63 500	46 700	-6.3%	-7.3%	-4.1%	-3.3%

TOTAL   - Total construction employees
BLACK   - Total construction black employees
COL/IND - Total construction coloured and indian employees
WHITE   - Total construction white employees

Source : SA Statistics (1992:7.13)

TABLE C4 — NON-AGRICULTURAL WAGES AND SALARIES

Excluding working proprietors

YEAR	NON-AGRICULTURAL WAGES & SALARIES			GROWTH ON PREVIOUS YEAR	
	ALL SECT	CONSTR.	CONSTR. %	ALL SECT	CONSTR.
	Thousands (R)				
1972	6 305 852	504 520	8.0%	–	
1973	7 374 557	641 302	8.7%	16.9%	27.1%
1974	8 916 870	833 699	9.3%	20.9%	30.0%
1975	10 727 066	1 003 404	9.4%	20.3%	20.4%
1976	12 430 373	1 047 724	8.4%	16.9%	4.4%
1977	13 560 631	957 574	7.1%	9.1%	-8.8%
1978	14 020 513	910 588	6.1%	8.3%	-4.9%
1979	16 927 147	977 692	5.8%	14.2%	7.4%
1980	20 901 894	1 305 756	6.2%	23.5%	33.8%
1981	26 250 262	1 721 605	6.6%	25.8%	31.8%
1982	31 645 833	2 151 224	6.8%	20.6%	25.0%
1983	35 343 339	2 464 678	7.0%	11.7%	14.6%
1984	41 815 213	2 725 488	6.5%	18.3%	10.6%
1985	46 222 790	2 970 133	6.4%	10.5%	9.0%
1986	53 179 860	3 218 979	6.1%	15.1%	8.4%
1987	61 885 388	3 514 722	5.7%	16.4%	9.2%
1988	72 449 916	4 249 218	5.9%	17.1%	20.9%
1989	85 863 046	4 696 147	5.5%	18.5%	10.5%
1990	100 206 367	5 235 012	5.2%	15.7%	11.4%
1991	113 957 888	5 808 003	5.1%	13.7%	10.9%

ALL SECT   — Total non agricultural wages & salaries
CONSTR   — Construction wages & salaries
CONSTR %   — Construction propration of total wages and salaries

Current prices
Source : SA Statistics (1990:7.8-7.13)

TABLE C5 — CONSTRUCTION WAGES AND SALARIES BY RACE GROUP

Excluding working proprietors

YEAR	TOTAL	BLACK	COL/IND	WHITE	GROWTH ON PREVIOUS YEAR			
					TOTAL	BLACK	COL/IND	WHITE
	Thousand (R)							
1972	504 520	173 722	91 013	239 785	–	–	–	–
1973	641 302	241 931	114 379	284 992	27.1%	39.3%	25.7%	18.9%
1974	833 699	334 321	149 363	350 015	30.0%	38.2%	30.6%	22.8%
1975	1 003 404	423 333	182 126	397 945	20.4%	26.6%	21.9%	13.7%
1976	1 047 724	439 213	198 148	410 364	4.4%	3.8%	8.8%	3.1%
1977	957 574	386 467	173 655	397 452	-0.8%	-12.0%	-12.4%	3.1%
1978	910 588	349 617	163 308	397 663	-4.9%	-9.5%	-5.0%	0.1%
1979	977 692	379 791	165 650	432 251	7.4%	8.6%	1.4%	8.7%
1980	1 305 756	550 159	237 324	518 273	33.5%	44.9%	43.3%	19.9%
1981	1 721 605	731 967	313 648	675 990	31.8%	33.0%	32.2%	30.4%
1982	2 151 224	927 048	402 474	821 702	25.0%	26.7%	28.3%	21.6%
1983	2 464 678	1 077 898	458 437	928 343	14.6%	16.3%	13.9%	13.0%
1984	2 725 488	1 199 928	508 357	1 017 203	10.6%	11.3%	10.9%	9.6%
1985	2 970 133	1 339 837	552 613	1 077 683	9.0%	11.7%	8.7%	6.9%
1986	3 218 979	1 438 642	585 664	1 195 673	8.4%	7.2%	6.2%	10.9%
1987	3 514 722	1 618 653	606 502	1 289 667	9.2%	12.7%	3.4%	7.9%
1988	4 249 216	1 987 683	711 039	1 550 514	20.9%	22.8%	17.2%	20.2%
1989	4 898 147	2 147 687	832 442	1 718 018	10.6%	8.1%	17.1%	10.8%
1990	5 235 012	2 502 634	879 895	1 852 483	11.4%	16.5%	5.7%	7.8%
1991	5 803 033	2 750 612	940 846	2 111 675	10.9%	9.9%	6.9%	14.0%

TOTAL   — Total of black, coloured and indian, and white
BLACK   — Total of black only
COL/IND   — Total of coloured and indian only
WHITE   — Total of white only

Current prices
Source : SA Statistics (1992: 7.13)

TABLE C6 - AVERAGE NON-AGRICULTURAL WAGES AND SALARIES BY RACE GROUP

Excluding working proprietors

YEAR	TOTAL		BLACK		COLOURED & INDIAN		WHITE	
	ALL SECT	CONSTR.	ALL SECT	CONSTR	ALL SECT	CONSTR	ALL SECT	CONSTR
	Rands per month							
1972	133	122	48	62	100	134	317	383
1973	146	132	57	72	115	148	350	411
1974	169	146	72	83	130	185	398	439
1975	198	173	91	103	150	200	450	505
1976	220	187	107	111	168	222	490	553
1977	246	212	121	123	185	230	535	635
1978	274	240	137	134	205	253	584	745
1979	306	259	156	147	226	249	654	817
1980	364	299	190	179	273	312	766	960
1981	437	346	230	207	332	373	933	1 138
1982	516	401	273	242	397	439	1 094	1 327
1983	584	480	311	295	455	524	1 213	1 520
1984	682	535	365	333	533	578	1 409	1 859
1985	765	604	-	384	-	662	-	1 806
1986	875	666	-	418	-	722	-	2 046
1987	1 003	719	-	480	-	759	-	2 244
1988	1 156	856	-	555	-	887	-	2 670
1989	1 360	938	-	592	-	1 037	-	2 976
1990	1 594	1 045	-	688	-	1 107	-	3 196
1991	1 847	1 237	-	816	-	1 235	-	3 768

ALL SECT - Average for all non-agricultural sectors including construction
CONSTR   - Average for construction sector only

Current prices
Source : SA Statistics (1992, 7.8 & 7.14)

TABLE C7 - GROWTH IN NON-AGRICULTURAL WAGES AND SALARIES BY RACE GROUP

Excluding working proprietors

YEAR	TOTAL		BLACK		COLOURED & INDIAN		WHITE	
	ALL SECT	CONSTR	ALL SECT	CONSTR	ALL SECT	CONSTR	ALL SECT	CONSTR
1972	-	-	-	-	-	-	-	-
1973	9.8%	8.2%	18.8%	16.1%	15.0%	10.4%	10.4%	7.3%
1974	15.8%	10.6%	26.3%	15.3%	13.0%	11.5%	13.7%	6.8%
1975	17.2%	18.5%	26.4%	24.1%	15.4%	21.2%	13.1%	15.0%
1976	11.1%	8.1%	17.6%	7.8%	12.0%	11.0%	8.9%	9.5%
1977	11.8%	13.4%	13.1%	10.8%	10.1%	3.6%	9.2%	14.8%
1978	11.4%	13.2%	13.2%	8.9%	10.8%	10.0%	9.2%	17.3%
1979	12.4%	7.9%	13.9%	9.7%	10.2%	-1.6%	12.0%	9.7%
1980	18.2%	15.4%	21.8%	21.8%	20.8%	25.3%	17.1%	17.5%
1981	20.1%	15.7%	21.1%	15.6%	21.6%	19.6%	21.8%	18.5%
1982	18.1%	15.9%	18.7%	16.9%	19.8%	17.7%	17.3%	16.6%
1983	13.2%	19.7%	13.9%	21.9%	14.6%	19.4%	10.9%	14.5%
1984	16.8%	11.5%	17.4%	12.9%	17.1%	10.3%	16.2%	9.1%
1985	12.2%	12.9%	-	15.3%	-	14.5%	-	8.9%
1986	14.4%	10.3%	-	8.9%	-	9.1%	-	13.3%
1987	14.6%	8.0%	-	10.0%	-	5.1%	-	9.7%
1988	15.3%	19.1%	-	20.7%	-	16.9%	-	19.0%
1989	17.6%	9.6%	-	6.7%	-	16.9%	-	11.5%
1990	17.2%	11.4%	-	16.2%	-	6.8%	-	7.4%
1991	15.9%	18.4%	-	18.6%	-	11.6%	-	17.9%

ALL SECT - Average for all non-agricultural sectors including construction
CONSTR   - Average for construction sector only

Current prices
Source : SA Statistics (1992)

TABLE C8 - INVESTMENT IN FIXED ASSETS AND INVENTORIES

YEAR	MACH/EQ	TRANS EQ	TRANSFER	BLDNGS	CIV ENG	TOTAL	BLD+CIV	BLD+CIV/ TOTAL	BUILD/ TOTAL
				Million (R)					
1983	8 347	2 932	692	6 691	4 613	23 275	11 304	48.6%	28.7%
1984	9 209	2 575	725	7 542	4 601	24 652	12 143	49.3%	30.6%
1985	11 608	2 860	634	7 884	5 187	28 173	13 074	46.4%	28.0%
1986	11 584	2 881	662	8 603	5 475	29 205	14 078	48.2%	29.5%
1987	12 576	3 662	928	9 174	5 723	32 063	14 897	46.5%	28.6%
1988	15 826	5 555	1 241	10 478	6 809	39 909	17 287	43.3%	26.3%
1989	18 762	7 069	1 380	12 844	8 584	48 639	21 428	44.1%	26.4%
1990	20 529	7 207	1 609	14 628	9 203	53 176	23 831	44.8%	27.5%
1991	20 217	8 422	1 870	14 890	8 558	53 957	23 448	43.5%	27.6%

MACH/EQ	- Machinery and equipment
TRANS EQ	- Transport equipment
TRANSFER	- Transfer costs
BLDNGS	- Buildings
CIV ENG	- Construction works
TOTAL	- Total investment
BLD+CIV	- Total of buildings and construction works
BLD+CIV/TOTAL	- Buildings and construction works as a percentage of total investment
BUILD/TOTAL	- Buildings as a percentage of total

Current prices
Source: BIFSA Annual report (1983-1988) and BIFSA Statistical Yearbook (1989-1992)

TABLE C9 — EMPLOYMENT IN THE CONSTRUCTION AND BUILDING INDUSTRIES

YEAR	BUILDING INDUSTRY +			CIVIL ENGINEERING INDUSTRY +			CONSTRUCTION INDUSTRY		
	WORK PROP	PD EMP	TOT1	WORK PROP	PD EMP	TOT2	WORK PROP	PD EMP	TOT3
						Employment numbers			
1972	4 983	244 507	249 490	187	99 716	99 803	5 170	344 883	349 305
1974	6 029	315 780	321 809	796	160 398	161 194	6 825	476 178	483 003
1976	7 229	286 436	293 665	611	179 973	180 584	7 840	466 409	474 249
1978	6 091	196 230	202 321	455	121 627	122 089	6 648	315 657	322 403
1980	6 151	230 513	236 664	362	133 632	133 994	6 513	364 145	370 658
1982	8 929	290 612	299 440	508	156 354	156 782	9 338	446 866	456 208
1985	9 922	261 703	271 625	244	149 397	149 341	10 865	410 130	420 305

WORK PROP — Working proprietors
PD EMP   — Paid employees
TOT1   — Total building industry working proprietors and paid employees
TOT2   — Total civil engineering industry working proprietors and paid employees
TOT3   — Total construction sector working proprietors and paid employees

Current prices
Source : Census of Construction (1985; 1-8)

TABLE C10 — EMPLOYMENT IN THE BUILDING INDUSTRY

YEAR	HOMEBUILDING +			GENERAL CONTRACTING +			SPECIAL CONTRACTING +			BUILDING INDUSTRY		
	WORK PROP	PD EMP	TOT1	WORK PROP	PD EMP	TOT2	WORK PROP	PD EMP	TOT3	WORK PROP	PD EMP	TOT4
									Employment numbers			
1972				2 824	176 586	178 769	2 659	68 122	70 781	4 983	244 507	
1974				2 668	229 564	233 204	3 394	95 216	98 600	6 029	315 780	
1976				3 096	196 585	199 665	4 133	89 847	93 980	7 229	286 436	
1978				3 396	123 141	128 571	3 781	75 049	78 750	6 091	194 230	
1980				2 461	161 306	163 767	3 870	79 207	82 877	6 151	230 513	
1982	3 394	95 821	102 215	811	99 744	99 855	5 823	102 047	107 870	8 828	289 812	
1985	3 971	75 666	79 637	271	97 674	97 945	6 780	95 353	104 143	9 922	261 703	

WORK PROP — Working proprietors
PD EMP   — Paid employees
TOT1   — Total homebuilding division working proprietors and paid employees
TOT2   — Total general contracting division working proprietors and paid employees
TOT3   — Total special contracting division working proprietors and paid employees
TOT4   — Total building industry working proprietors and paid employees

Shaded area indicates total of homebuilding and general contracting
Current prices
Source : Census of Construction (1985; 1-8)

TABLE C11 - BUILDING INDUSTRY EMPLOYMENT PROPORTIONAL
TO CONSTRUCTION INDUSTRY EMPLOYMENT

YEAR	WORK PROP	PD EMPL	TOTAL
1972	96.4%	71.0%	71.4%
1974	88.3%	66.3%	66.6%
1976	92.2%	61.4%	61.9%
1978	93.0%	61.5%	62.1%
1980	94.4%	63.3%	63.8%
1982	94.6%	65.0%	65.6%
1985	91.3%	63.8%	64.5%

WORK PROP — Working proprietors
PD EMPL   — Paid employees
TOTAL   — Working proprietors and paid employees

Source : Census of Construction (1985: 1–2)

TABLE C12 - EMPLOYMENT IN THE GENERAL CONTRACTING DIVISION OF THE BUILDING INDUSTRY

YEAR	PROPORTION OF CONSTRUCTION			PROPORTION OF BUILDING		
	WORK PROP	PD EMPL	TOTAL1	WORK PROP	PD EMPL	TOTAL2
1972						
1974						
1976						
1978						
1980						
1982	2.3%	20.1%	19.7%	2.4%	30.9%	30.0%
1985	2.5%	21.4%	20.9%	2.7%	33.5%	32.4%

WORK PROP — Working proprietors
PD EMPL   — Paid employees
TOTAL1   — Working proprietors and paid employees proportion of construction industry
TOTAL2   — Working proprietors and paid employees proportion of building industry

Source : Census of Construction (1985: 1–2)

TABLE C13 - EMPLOYMENT IN THE HOMEBUILDING DIVISION OF THE BUILDING INDUSTRY

YEAR	PROPORTION OF CONSTRUCTION			PROPORTION OF BUILDING		
	WORK PROP	PD EMPL	TOTAL1	WORK PROP	PD EMPL	TOTAL2
1972						
1974						
1976						
1978						
1980						
1982	36.4%	22.1%	22.4%	38.4%	34.0%	34.1%
1985	35.6%	18.5%	18.9%	39.0%	28.9%	29.3%

See notes for Table C12

TABLE C14 - EMPLOYMENT IN THE SPECIAL CONTRACTING DIVISION OF THE BUILDING INDUSTRY

YEAR	PROPORTION OF CONSTRUCTION			PROPORTION OF BUILDING		
	WORK PROP	PD EMPL	TOTAL1	WORK PROP	PD EMPL	TOTAL2
1972	51.4%	19.8%	20.3%	53.4%	27.9%	28.4%
1974	49.6%	20.0%	20.4%	56.1%	30.2%	30.6%
1976	52.7%	19.3%	19.6%	57.2%	31.4%	32.0%
1978	56.5%	23.1%	23.8%	60.8%	37.6%	38.3%
1980	56.3%	21.8%	22.4%	59.7%	34.4%	35.0%
1982	55.9%	22.8%	23.5%	59.2%	35.1%	35.8%
1985	53.2%	24.0%	24.7%	58.3%	37.6%	38.3%

See notes for Table C12

TABLE C15 – GROSS OUTPUT IN THE CONSTRUCTION AND BUILDING INDUSTRIES

YEAR	CONSTRUCTION	TOT BLD	CIV ENG	HOMEBUILD	GEN CONTR	SPEC CONTR
			Thousand (R)			
1972	1 400 987	1 051 114	399 873		649 453	351 661
1974	2 351 539	1 559 773	791 766		962 862	576 911
1976	3 222 375	1 970 608	1 251 766		1 180 824	789 982
1978	2 832 486	1 725 509	1 106 977		928 177	800 332
1980	3 765 997	2 223 702	1 537 295		1 237 284	984 418
1982	6 986 945	4 222 491	2 764 454	1 325 603	1 035 928	1 860 960
1985	9 961 001	6 373 408	3 587 582	1 239 864	1 763 787	2 849 758

CONSTRUCTION – Gross output of construction sector
TOT BLD   – Gross output of building industry
CIV ENG   – Gross output of civil engineering industry
HOMEBUILD   – Gross output of homebuilding division of building industry
GEN CONTR   – Gross output of general contracting division of building industry
SPEC CONTR   – Gross output of special contracting division of building industry

Current prices
Source : Census of Construction (1985 1-8)

TABLE C16 – PROPORTIONAL BREAKDOWN OF GROSS OUTPUT

YEAR	BUILDING/ CONSTR	CIV ENG/ CONSTR	HOMEBUILD/ CONSTR	HOMEBUILD/ BUILD	SPEC CON/ CONSTR	SPEC CON/ BUILD	GEN CON/ CONSTR	GEN CON-HO/ CONSTR	GEN CON/ BUILD	GEN CON+HOME/ BUILD
1972	71.5%	28.5%			25.1%	35.1%	46.4%	46.4%		64.9%
1974	66.0%	34.0%			24.7%	37.5%	41.3%	41.3%		62.5%
1976	61.2%	38.8%			24.5%	40.1%	36.8%	36.6%		59.9%
1978	60.9%	39.1%			26.4%	46.7%	32.5%	32.5%		53.3%
1980	59.1%	40.9%			26.2%	44.4%	38.9%	32.9%		55.6%
1982	60.4%	39.6%	19.0%	31.4%	25.6%	44.1%	14.6%	33.8%	24.5%	55.9%
1985	64.0%	36.0%	17.5%	37.3%	28.6%	45.0%	17.7%	35.3%	27.7%	55.0%

BUILDING/CONSTR.   – Building industry gross output as a proportion of construction industry gross output
CIV ENG/CONSTR   – Civil engineering industry gross output as a proportion of construction industry gross output
HOMEBUILD/CONSTR   – Homebuilding division gross output as a proportion of construction industry gross output
HOMEBUILD/BUILD   – Homebuilding division gross output as a proportion of building industry gross output
SPEC CON/CONSTR   – Special contracting division gross output as a proportion of construction industry gross output
SPEC CON/BUILD   – Special contracting division gross output as a proportion of building industry gross output
GEN CON/CONSTR   – General contracting division gross output as a proportion of construction industry gross output
GEN CON/BUILD   – General contracting division gross output as a proportion of building industry gross output
GEN CON+HOME   – General contracting and homebuilding

Shaded areas indicate no data available
Source : Census of Construction (1985 1-8)

TABLE C17 - GROSS DOMESTIC FIXED INVESTMENT IN BUILDINGS - PUBLIC AND PRIVATE SECTOR

YEAR	PUBLIC SECTOR +			PRIVATE SECTOR =			GRAND TOTAL		
	RES BUILD	NON-RES BLD	TOTAL1	RES BUILD	NON-RES BLD	TOTAL2	RES BUILD	NON-RES BLD	TOTAL3
				Millions (R)					
1980	1 093	1 931	3 024	2 200	1 782	3 982	3 293	3 713	7 006
1981	1 078	1 961	3 039	2 621	2 280	4 901	3 699	4 241	7 940
1982	939	1 748	2 687	2 926	2 723	5 649	3 865	4 471	8 336
1983	1 019	1 643	2 662	3 126	2 475	5 601	4 145	4 118	8 263
1984	1 029	1 786	2 815	3 567	2 145	5 712	4 596	3 931	8 527
1985	838	1 893	2 731	3 013	2 320	5 333	3 851	4 213	8 064
1986	649	1 716	2 365	2 697	2 016	4 713	3 346	3 732	7 078
1987	584	1 577	2 161	2 760	1 611	4 371	3 344	3 188	6 532
1988	580	1 479	2 059	2 867	1 905	4 772	3 447	3 384	6 831
1989	593	1 546	2 139	2 835	2 292	5 127	3 428	3 838	7 266
1990	514	1 820	2 334	2 568	2 403	4 971	3 082	4 223	7 305
1991	541	1 317	1 858	2 436	2 376	4 812	2 977	3 693	6 670
1992	525	1 010	1 535	2 327	2 251	4 578	2 852	3 261	6 113

RES BUILD   - Residential buildings
NON-RES BLD - Non-residential buildings
TOTAL1   - Public sector total of residential and non-residential
TOTAL2   - Private sector total of residential and non-residential

Constant (1985) prices
Source : BIFSA Statistical yearbook (1992 3-4)

TABLE C18 - GROSS DOMESTIC FIXED INVESTMENT IN BUILDINGS - PROPORTIONAL CONTRIBUTION OF PUBLIC AND PRIVATE SECTORS

YEAR	PUBLIC SECTOR			PRIVATE SECTOR		
	RES BUILD	NON-RES BLD	TOTAL1	RES BUILD	NON-RES BLD	TOTAL2
1980	33.2%	52.0%	43.2%	66.8%	48.0%	56.8%
1981	29.1%	46.2%	38.3%	70.9%	53.8%	61.7%
1982	24.3%	39.1%	32.2%	75.7%	60.9%	67.8%
1983	24.6%	39.9%	32.2%	75.4%	60.1%	67.8%
1984	22.4%	45.4%	33.0%	77.6%	54.6%	67.0%
1985	21.8%	44.9%	33.9%	78.2%	55.1%	66.1%
1986	19.4%	46.0%	33.4%	80.6%	54.0%	66.6%
1987	17.5%	49.5%	33.1%	82.5%	50.5%	66.9%
1988	16.8%	43.7%	30.1%	83.2%	56.3%	69.9%
1989	17.3%	40.3%	29.4%	82.7%	59.7%	70.6%
1990	16.7%	43.1%	32.0%	83.3%	56.9%	68.0%
1991	18.2%	35.7%	27.9%	81.8%	64.3%	72.1%
1992	18.4%	31.0%	25.1%	81.6%	69.0%	74.9%

RES BUILD   - Residential buildings
NON-RES BLD - Non-residential buildings
TOTAL1   - Public sector proportion of all buildings
TOTAL2   - Private sector proportion of all buildings

Constant (1985) prices
Source : BIFSA Statistical yearbook (1992 3-4)

TABLE C19 - BLACK BUILDING INDUSTRY EMPLOYEES BY OCCUPATION - 1985

DIVISION	HOMEBUILDING			GENERAL CONTRACTING			SPECIAL CONTRACTING		
OCC	TOTAL	BLACK	BLACK P	TOTAL	BLACK	BLACK P	TOTAL	BLACK	BLACK P
		Number	%		Number	%		Number	%
FMEN	2 635	151	5.7%	2 961	115	3.9%	3 407	235	6.9%
ART/APP	11 174	2 078	18.6%	12 064	2 633	21.8%	14 757	2 402	16.3%
PR WK/OP	5 060	3 321	65.6%	9 071	7 372	81.3%	9 377	5 931	63.3%
WORKERS	51 618	40 251	78.0%	58 053	53 505	92.2%	60 483	50 159	82.9%

OCC   - Occupation
FMEN   - Foreman
ART/APP   - Artisans and apprentices
PR WK/OP   - Production workers and operators
BLACK P - Blacks as a proportion of total

Source : Census of Construction (1985: 74-75)

APPENDIX D

D1 - MODEL FOR THE DEVELOPMENT OF SMALL-SCALE BUILDERS
(as designed by the Development Bank of Southern Africa) (Calcopietro, 1992)

** ENTREPRENEURS CAN BE IDENTIFIED IN A CONTINUUM OF DEVELOPMENT FROM "EMERGING"*
TO "ESTABLISHED"

Continuum	Managerial competencies expected
Specified trade skills/ New entrant	A prospective employee with few specific skills in any of the trades; bricklaying; carpentry; plastering; tiling; painting; or plumbing and capable of performing menial tasks or acting as an artisan's assistant.
Trade skills/Artisan	The individual is capable of piece work using his own trade skills (for example, a bricklayer)
Emerging Small Contractor/Sub-contractor	A sub-contractor who could be employed by an established contractor, practising his trade skill and employing his own worker(s). A contractor who could manage a single unit house building or alterations contract and employs either his own workers or sub contractors. All materials could be supplied and paid for by the client.
Established Small Contractor	An entrepreneur who operates multi-unit medium sized contracts for his own account and is usually responsible for both labour and material. Employs sub-contractors and his own employees, should also be involved in marketing and is likely to operate in the informal sector of the industry.
Emerging Contractor	An entrepreneur who operates several contracts which could include domestic and commercial work. He would also tender on the open market and should therefore be able to operate in the formal sector.

Appendix D   Page D1

<u>**D2 - SUPPORT SERVICES FOR THE DEVELOPMENT OF EMERGING CONTRACTORS**</u>
(as recommended by the Development Bank of Southern Africa) (Calcopietro, 1992)

SUPPORT SERVICES USUALLY REDUCE AS THE CAPABILITY LEVEL OF THE CONTRACTOR INCREASES, AND NORMALLY INCLUDE :

Access to finance

In low-income communities, where financial resources are limited, appropriate financial support is essential to assist the emerging contractors with their endeavours. Consideration should be given to:

* *short term bridging finance;*
* *medium term finance for working capital and equipment;*
* *long term finance for physical facilities or buildings.*

Technical and Management Counselling and Training

Will contribute towards the development of the local construction industry. This should not be limited to the emerging contractors, but should include communities and interested individuals.

Access to reasonably priced building materials

The cost of materials often constitutes the largest proportion of total project cost. The implementing agency can play a facilitatory role in achieving reasonably priced materials for the emerging contractor, either by :

* *supplying directly;*
* *organising discounts through existing outlets; or*
* *providing adequate finance for bulk purchase of materials.*

Design and estimating

The design of the project is the most important factor to consider when providing affordable and acceptable facilities for implementation by emerging contractors. Professional consultants who understand the approach to emerging contractors should be appointed by the implementing agency. Professional fees are usually capitalised against project cost.

Construction services

The implementing agency should avoid acting as the main contractor itself, especially if this function can be fulfilled by emerging contractors.

THE SURVEY QUESTIONNAIRE

QUESTIONNAIRE

MODULE A - SURVEY INFORMATION

A_1   Survey number   ________________

A_2   Interviewer's number   ________________

A_3   CITY   (1) Durban   (2) Johannesburg

A_4   DATE   ________________(YY/MM/DD)

A_5   Type of firm

        A_5_1 General contractor
                (takes on whole building - employer holds him responsible for entire end product)

                1)   mainly new work
                2)   mainly alterations & additions work

        A_5_2 Specialist sub-contractor
                (provides labour & material)

                1)   electrical
                2)   plumbing & drainage

        A_5_3 Labour only sub-contractor

                1)   brickwork/blockwork
                2)   carpentry (roof timbers & sheet roof coverings)
                3)   plastering (plaster & screeds)
                4)   painting
                5)   ceilings
                6)   combination of the above

        A_5_4 Labour and material sub-contractor

                1)   combination of trades

MODULE B - GENERAL

B_1   When did you start this firm?

        B_1_1 Year
        B_1_2 Month

B_2   Give the name of the builders association/s that your firm belongs to

    B_2_A ____

    B_2_B ____

B_3   Is this the first business you have started?

    1) Yes
    2) No

B_4   Are you the sole owner of your business?

    1) Yes
    2) No

    B_4_1   If no, how many partners do you have?

MODULE C - PERSONAL BACKGROUND OF OWNER

C_1   Age

C_2   What is/was your father's occupation?

    1)   building industry employee/tradesman
    2)   businessman
    3)   other

C_3   What is the highest standard you passed at school?

1)	Standard 1	2)	Standard 2
3)	Standard 3	4)	Standard 4
5)	Standard 5	6)	Standard 6
7)	Standard 7	8)	Standard 8
9)	Standard 9	10)	Standard 10
11)	Sub A	12)	Sub B

C_4   Have you obtained a qualification from a college, Technikon, or university?

    1)   Yes
    2)   No

    C_4_1   If yes, what type of qualification have you obtained from a college, Technikon, or university?

    ____

C_5   Is building your main source of income?

    1)   Yes
    2)   No

C_6   Indicate which languages you speak, read and write best

1) Afrikaans   2) English   3) Matabele   4) Portuguese   5) SeSotho   6) Shangaan
7) Swazi   8) Tsonga   9) Tswana   10) Xhosa   11) Zulu

C_6_1 _____________ (first most important)
C_6_2 _____________ (second most important)

MODULE D - TRAINING

D_1   Were you an employee in the building industry prior to staring this firm?

1)   Yes
2)   No

D_1_1 If yes, in what capacity and for how long?

Multiple response

D_1_1_1	Unskilled	_____________ (months)
D_1_1_2	Semi-skilled	_____________ (months)
D_1_1_3	Skilled	_____________ (months)
D_1_1_4	Foreman	_____________ (months)
D_1_1_5	Clerical	_____________ (months)
D_1_1_6	Management	_____________ (months)
D_1_1_X	Other (specify)	_____________ (months)

D_2   Have you served a formal apprenticeship?

1)   Yes
2)   No

If yes, in what trade or trades?

Multiple response

D_2_1_1	Bricklaying	1) Yes
D_2_1_2	Plastering	1) Yes
D_2_1_3	Plumbing	1) Yes
D_2_1_4	Electrical	1) Yes
D_2_1_5	Carpentry	1) Yes
D_2_1_6	Painting	1) Yes
D_2_1	Other (specify)	

D_3   Did you receive any formal education or training in how to manage a building firm prior to starting this business?

1)   Yes
2)   No

D_4 Do you feel that you require further training?

    1)   Yes
    2)   No

    D4_1   If yes, in which of the following fields?

		Multiple response
D_4_1_1	Skills training	1) Yes
D_4_1_2	Production management	1) Yes
	(management of the building process on site)	
D_4_1_3	Business management	1) Yes
	(financial management, job coordination, administration, *etc.*)	
D_4_1_4X	Other	

    D_4_1_5X _____________________________

    D_4_2 What have you done about getting this training?

        1)   Nothing
        2)   Something

    D_4_3 If you have done something, then explain the outcome

    ____

    D_4_4 Who would you go to for this training?
        (Give the name of the institution/s if your choice, leave blank if 'don't know')

        D_4_4A_______________________________

        D_4_4B_______________________________

MODULE E - WORK HISTORY

The interviewer must ask 8 questions.

Question 1   What were you doing in January to March 1990?

    Then respondent must choose only one of the six activities identified as E_1_1 to E_1_6

Question 2   What were you doing in April to June 1990?

    Then respondent must choose only one of the six activities identified as E_1_7 to E_1_12, *etc.*

1) Yes

Period	Current firm had work	Firm assisted another firm	Reverted to employee status	Not active - sought work	Left the building industry	None of these
	E_1_1	E_1_2	E_1_3	E_1_4	E_1_5	E_1_6
Jan-Mar 1990						
	E_1_7	E_1_8	E_1_9	E-1-10	E_1_11	E_1_12
Apr-Jun 1990						
	E_1_13	E_1_14	E_1_15	E_1_16	E_1_17	E_1_18
Jul-Sept 1990						
	E_1_19	E_1_20	E_1_21	E_1_22	E_1_23	E_1_24
Oct-Dec 1990						
	E_1_25	E_1_26	E_1_27	E_1_28	E_1_29	E_1_30
Jan-Mar 1991						
	E_1_31	E_1_32	E_1_33	E_1_34	E_1_35	E_1_36
Apr-Jun 1991						
	E_1_37	E_1_38	E_1_39	E_1_40	E_1_41	E_1_41
Jul-Sept 1991						
	E_1_43	E_1_44	E_1_45	E_1_46	E_1_47	E_1_48
Oct-Dec 1991						

MODULE F - ENTRY INTO INDUSTRY

F_1   What motivated you to start your current firm?

1) Family business/inherited
2) Unemployed/difficult to find job
3) Independence
4) Thought you could earn more than in another job
5) Lack of education
6) Ambition
7) Wanted to leave sub-contracting

F_2   Did your previous employer help you to start your current firm?

1) Yes
2) No

<u>**N.B. IF 'NO' THEN MOVE TO QUESTION F_4**</u>

F_2_1 If 'yes', what was your previous employer's business?

    1) Building industry
    2) Other industry

F_2_2 How did your previous employer help you to start?

Multiple response

F_2_2_1	Loan	1) Yes
F_2_2_2	Passed on excess work	1) Yes
F_2_2_3	Gave you contacts which resulted in work	1) Yes
F_2_2_X	Other (specify)	

____

F_3 Do you still receive any help from your previous employer?

    1) Yes
    2) No

<u>**N.B. IF 'NO' THEN MOVE TO QUESTION F_4**</u>

F_3_1 If 'yes', what kind of help?

Multiple response

F_3_1_1	Loan	1) Yes
F_3_1_2	Passed on excess work	1) Yes
F_3_1_3	Gave you contacts which resulted in work	1) Yes
F_3_1_X	Other (specify)	

____

F_4 If your previous employer did not get you your first contract, how did you get it?

Explain:

F_4A _______________________________

F_4B _______________________________

F_5 List the two most serious problems you experienced in starting your own business

F_5_1 _______________________________

F_5_2 _______________________________

MODULE G - RECENT & CURRENT ACTIVITIES OF FIRM

<u>N.B.</u>   <u>QUESTION G_1 MUST ONLY BE PUT TO GENERAL CONTRACTORS</u>

G_1   If you are a **general contractor**, how often were you employed by the following type of employer since t
beginning of 1990?

1) Never   2) Sometimes   3) Always

G_1_1   Individuals (home owners)
G_1_2   Organisations
G_1X   Other (specify)

____

<u>N.B.</u>   <u>QUESTION G_2 MUST ONLY BE PUT TO SUB-CONTRACTORS</u>

G_2   If you are a **sub-contractor**, how often were you employed by the following type of employer since the
beginning of 1990?

1) Never   2) Sometimes   3) Always

G_2_1   Developers
G_2_2   Large contractors
G_2_3   Small and medium contractors
G_2_4   Individuals (building owners)
G_2X   Other (specify)

____

G_3   How often do you obtain work by the following methods?

1) Never   2) Sometimes   3) Always

G_3_1   Competitive tender
G_3_2   Negotiation
G_3_3   "Take-it-or-leave-it" offer by employer
G_3X   Other, (specify)

____

G_4   Number of contracts/jobs done by your firm in the last year (1991)?

___________________ (raw number)

GIVE THE FOLLOWING DETAILS OF YOUR LAST (MOST RECENT) CONTRACT

G_5   Type of employer?

    1)   Owner
    2)   Developer
    3)   SAHT/HODECO/UTILITY CO. (service organisations)

G_5_1 Type of building?

    1)   Commercial
    2)   Residential
    3)   Service

G_5_2 What did your company actually do?

Multiple response

G_5_2_1	Brickwork	1) Yes
G_5_2_2	Plastering	1) Yes
G_5_2_3	Painting	1) Yes
G_5_2_4	Plumbing	1) Yes
G_5_2_5	Electrical wiring 1) Yes	
G_5_2_6	Carpentry	1) Yes
G_5_2X	Other (specify)	

____

G_5_3 What was ___________ ?

    G_5_3_1   Agreed contract value   R___________

    G_5_3_2   Actually received   R___________

G_5_4 What was the contract period?   ___________ (months)

G_5_5 What were the following costs?

    G_5_5_1 Material   ___________ (Rand value)

    G_5_5_2 Labour   ___________ (Rand value)

    G_5_5_3 Equipment hire   ___________ (Rand value)

    G_5_5_4 Other (excluding
    owners salaries)   ___________ (Rand value)

G_5_6 What salary did you draw from this job for yourself and your partners ___________ (Rand value)

G_5_7 What was your profit?   ___________ (Rand value)

GIVE THE FOLLOWING DETAILS OF YOUR FIRST CONTRACT

G_6   Type of employer?

    1)   Owner
    2)   Developer
    3)   SAHT/HODECO/UTILITY CO. (service organizations)

G_6_1   Type of building?

    1)   Commercial
    2)   Residential
    3)   Service

G_6_2   What did your company actually do?

Multiple response

G_6_2_1	Brickwork	1) Yes
G_6_2_2	Plastering	1) Yes
G_6_2_3	Painting	1) Yes
G_6_2_4	Plumbing	1) Yes
G_6_2_5	Electrical wiring	1) Yes
G_6_2_6	Carpentry	1) Yes
G_6_2X	Other (specify)	

____

G_6_3   What was _____________?

    G_6_3_1   Agreed contract value   R____________

    G_6_3_2   Actually received   R____________

G_6_4   What was the contract period?   _____________ (months)

G_6_5   What were the following costs?

    G_6_5_1 Material   _____________ (Rand value)

    G_6_5_2 Labour   _____________ (Rand value)

    G_6_5_3 Equipment hire   _____________ (Rand value)

    G_6_5_4 Other (excluding
            owners salaries)   _____________ (Rand value)

G_6_6   What salary did you draw from this job for yourself and your partners __________ (Rand value)

G_6_7   What was your profit?   _______________ (Rand value)

G_7   In which areas do you most often find work?

   1) Never   2) Sometimes   3) Always

   G_7_1   Own community
   G_7_2   Townships
   G_7_3   City
   G_7X   Other (specify)

   ____

G_8   Exactly what is your firm doing now?

   G_8_1   LABOUR

      1) Laid off
      2) Being paid, but not working
      3) Doing piecemeal work
      4) Working on site

   G_8_2   PLANT

      1) Returned to hirer
      2) Standing idle
      3) Hired out
      4) Used for piecemeal work
      5) On site

   G_8_3   OWNER

      1) Not active
      2) Negotiating new contracts
      3) Following-up debtors
      4) Doing alternative work
      5) Managing current job

G_9   How do you measure results regarding the success of your business?

   Explain:

   G_9A   _______________________________

   G_9B   _______________________________

MODULE H - FUTURE OBJECTIVES

H_1   What are your plans for your firm in the future?

      1)   Remain the same size
      2)   Get smaller (reduce no. jobs per year)
      3)   Get bigger (expand)
      4)   Sell the firm and leave industry
      5)   Sell the firm and seek better wage/salary elsewhere in building industry
      6)   Start a new business in the building industry
      7)   Diversify

H_2   If you plan to **expand**, what type of client/employer do you intend to get most of your work from?

      1)   Private families/individuals
      2)   Contractors
      3)   Principals
      4)   Provincial or government agencies

H_3   What steps have you taken/are you taking to achieve your objective?

            Multiple response

      H_3_1 Tried to diversify   1) Yes
      H_3_2 Advertised   1) Yes
      H_3_3 Invested in additional capital   1) Yes
      H_3_4 Approached possible principals   1) Yes
      H_3_5 Invested/saving   1) Yes

N.B.   **QUESTION H_4 MUST ONLY BE PUT TO GENERAL CONTRACTORS**

H_4   If you mainly do alteration & addition work, which of the following options would you choose?

      1)   Keep on doing alteration & addition work
      2)   Become a labour-only sub-contractor
      3)   Become a plumber
      4)   Become an electrician
      5)   Become a general contractor doing mainly new work
      6)   None of these

N.B.   **QUESTION H_5 MUST ONLY BE PUT TO SUB-CONTRACTORS**

H_5   If you are a sub-contractor, which of the following options would you choose?

      1)   Keep on doing alteration & addition work
      2)   Become a labour-only sub-contractor
      3)   Become a plumber
      4)   Become an electrician
      5)   Become a general contractor doing mainly new work
      6)   None of these

MODULE I - MANAGEMENT/ORGANISATION OF PRODUCTION

I_1 How do you obtain plant/equipment? (examples of plant: scaffolding, concrete mixers, compressors, drills)

Multiple response

I_1_1	Own	1) Yes
I_1_2	Hire	1) Yes
I_1_3	Lease	1) Yes
I_1X	Other (specify)	

I_2 Which of the following provide you with technical support?

I_2_1 Joint venture partner
I_2_2 Plant/equipment suppliers
I_2_3 Government/SBDC/CSIR, *etc.*
I_2_4 Industry association
I_2_5 General contractor who employed you
I_2_6 Materials suppliers
I_2_7 Material producers/manufacturers
I_2_8 NGOs
I_2_9 Private consultants

I_3 What type of bank account do you operate?

1) Separate bank accounts for personal and business transactions
2) One bank account for personal and business transactions
3) No bank account

I_4 Do you keep accurate financial records for your business?

1) Yes
2) No

I_5 Do you experience problems calculating a tender/quote amount?

1) Yes
2) No

I_6 How do you normally arrive at your tender/quote amount?

1) Calculate rate per square metre
2) Calculate quantities and costs of everything
3) Find out competitors costs and charge less

I_7   Have you ever felt you had to leave a site before completing your work?

       1) Yes
       2) No

       If 'yes', why?

I_7_1	Not paid	1) Yes
I_7_2	Ran out of finance	1) Yes
I_7_3	Forced out because of threats	1) Yes
I_7_4	General violence in area	1) Yes
I_7X	Other (specify)	

       ____

I_8   Do you calculate working capital requirements?

       1) Yes
       2) No

       I_8_1   If 'yes', how often?

              1)   One week in advance
              2)   One month in advance
              3)   Three months in advance
              4)   One year in advance

I_9   How many contracts/jobs can you do simultaneously before you experience coordination problems?

       _______________ (raw number)

MODULE J - CAPITAL AND FINANCING

J_1   Which of the following have been major sources of finance?

       1) Yes

		A Current	B At start-up
J_1_1	Own savings		
J_1_2	Family and friends		
J_1_3	Partners		
J_1_4	Informal market (money lenders)		
J_1_5	Formal commercial bank		
J_1_6	Government agency (e.g. SBDC)		
J_1_7	Welfare, churches and other donors		
J_1_8	Credit from main contractor		
J_1_9	Credit from material/plant suppliers		
J_1_10	Advance payment from employer		
J_1_11	Development corporation		

J_2   Have you experienced problems meeting security/collateral requirements when applying for loans?

   1) Yes
   2) No
   3) Not applicable (do not apply for loans)

   J_2_1   If 'yes', what were the problems?

J_2_1_1	Do not own property	1) Yes
J_2_1_2	Do not have capital	1) Yes
J_2_1_3	Do not have contracts to cede	1) Yes
J_2_1_4	Institutions are inaccessible	1) Yes
J_2_1_5	Have not been in business long enough	1) Yes
J_2_1X	Other (specify)	

   ____

J_3   Do you experience problems with loan repayment, fluctuating interest rates, etc?

   1) Yes
   2) No

J_4   Have you ever had a legal judgement against you?

   1) Yes
   2) No

   J_4_1   If 'yes', how has this affected your current business?

   Explain:

   ____

J_5   Give the following details of your money and investments at start-up and currently

		A Current	B Start-up
J_5_1	Cash in hand	Rand _________	_________
J_5_2	Hand tools	Rand _________	_________
J_5_3	Transport	Rand _________	_________
J_5_4	Electrically operated tools	Rand _________	_________
J_5·5	Petrol/diesel driven tools	Rand _________	_________
J_5X	Other	Rand _________	_________

MODULE K - LABOUR

K_1   Number of employees at start-up?

Type of employee	Full-time	Part-time	Casual	Family
	K_1_1_1	K_1_2_1	K_1_3_1	K_1_4_1
Cleaners				
	K_1_1_2	K_1_2_2	K_1_3_2	K_1_4_2
Labourers				
	K_1_1_3	K_1_2_3	K_1_3_3	K_1_4_3
Qualified artisans				
	K_1_1_4	K_1_2_4	K_1_3_4	K_1_4_4
Foremen				
	K_1_1_5	K_1_2_5	K_1_3_5	K_1_4_5
Apprentices				
	K_1_1_6	K_1_2_6	K_1_3_6	K_1_4_6
Clerks				

K_2   Number of employees at end of 1991?

Type of employee	Full-time	Part-time	Casual	Family
	K_2_1_1	K_2_2_1	K_2_3_1	K_2_4_1
Cleaners				
	K_2_1_2	K_2_2_2	K_2_3_2	K_2_4_2
Labourers				
	K_2_1_3	K_2_2_3	K_2_3_3	K_2_4_3
Qualified artisans				
	K_2_1_4	K_2_2_4	K_2_3_4	K_2_4_4
Foremen				
	K_2_1_5	K_2_2_5	K_2_3_5	K_2_4_5
Apprentices				
	K_2_1_6	K_2_2_6	K_2_3_6	K_2_4_6
Clerks				

K_3   Do you use casual labour?

    1) Always
    2) Sometimes
    3) Never

    K_3_1 On average, how many casuals do you employ at any one time?

N.B.   <u>QUESTION K_4 MUST ONLY BE PUT TO GENERAL CONTRACTORS</u>

K_4   If you are a general contractor, which of the following types of work do you sub-contract out?

    1) Always   2) Sometimes   3) Never

Type of work	Labour only	Labour and material
	K_4_1_1	K_4_2_1
Brickwork		
	K_4_1_2	K_4_2_2
Roof carpentry		
	K_4_1_3	K_4_2_3
Roof tiling		
	K_4_1_4	K_4_2_4
Plastering and screeds		
	K_4_1_5	K_4_2_5
Painting		
	K_4_1_6	K_4_2_6
Ceilings		
	K_4_1_7	K_4_2_7
Glazing		
	K_4_1_8	K_4_2_8
Floor coverings		
	K_4_1_9	K_4_2_9
Plumbing		
	K_4_1_10	K_4_2_10
Electrical		

K_5   How much did you pay your employees during 1991?

Type of employee	Amount in Rand	Specify frequency 1) Per MONTH 2) Per WEEK 3) Per DAY 4) Per Hour
	K_5_1_1	K_5_2_1
Cleaners		
	K_5_1_2	K_5_2_2
Labourers		
	K_5_1_3	K_5_2_3
Qualified artisans		
	K_5_1_4	K_5_2_4
Foremen		
	K_5_1_5	K_5_2_5
Apprentices		
	K_5_1_6	K_5_2_6
Clerks		

K_6   Is your firm registered with the Industrial Council?

    1) Yes
    2) No

    K_6_1 If 'no', why not?

    Explain:

____

K_7   How effectively can industrial council inspectors check on you?

    1) Not at all
    2) With difficulty
    3) Easily

MODULE L - CONSTRAINTS

L_1   Rank each of the following problems on a scale of 1 (lowest) to 5 (highest)

Circle the response

LABOUR

L_1_1	Cost of labour	1	2	3	4	5
L_1_2	Reliability of labour	1	2	3	4	5
L_1_3	Shortage of skilled labour	1	2	3	4	5
L_1_4	Relationship with main contractor	1	2	3	4	5
L_1_5	Relationship with sub-contractor	1	2	3	4	5
L_1_6	Worker action	1	2	3	4	5

FINANCE

L_1_7	Access to loans/security	1	2	3	4	5
L_1_8	Repayment of loans	1	2	3	4	5
L_1_9	Interest rate issues	1	2	3	4	5

PRODUCTION/MANAGEMENT

L_1_10	Lack technical skills	1	2	3	4	5
L_1_11	Supply/cost of materials	1	2	3	4	5
L_1_12	Cost of/access to equipment	1	2	3	4	5
L_1_13	Tender/negotiation procedures	1	2	3	4	5
L_1_14	Working out total costs	1	2	3	4	5
L_1_15	Accounting/reconciliation	1	2	3	4	5
L_1_16	Cash flow/slow payment	1	2	3	4	5

LEGISLATION/GOVERNMENT

L_1_17	Access to land	1	2	3	4	5
L_1_18	Tax legislation	1	2	3	4	5
L_1_19	Industrial councils/labour legislation	1	2	3	4	5
L_1_20	Other legislation	1	2	3	4	5

MARKETS

L_1_21	Preference for white contractors	1	2	3	4	5
L_1_22	Erratic inflow of work	1	2	3	4	5
L_1_23	Competition	1	2	3	4	5

Choose the FIVE most important constraints from those that you rated '4' or '5' in the previous question an
give details of the nature of the problem and what you feel could be done to eliminate it.

L_2   Problem A (state code from question L_1) L_   _________ [L_2_A]   _________ [L_2_B]

L_2_1   Nature of your problem A?

        Explain :

        ____

L_2_2   What would improve the situation relating to problem A?

        Explain:

        ____

L_3   Problem B (state code from question L_1) L_   _________ [L_3_A]   _________ [L_3_B]

L_3_1   Nature of your problem B?

        Explain :

        ____

L_3_2   What would improve the situation relating to problem B?

        Explain:

        ____

L_4   Problem C (state code from question L_1) L_   _________ [L_4_A]   _________ [L_4_B]

L_4_1   Nature of your problem C?

        Explain :

        ____

L_4_2 What would improve the situation relating to problem C?

Explain:

____

L_5 Problem D (state code from question L_1) L_ ___________ [L_5_A] _________ [L_5_B]

L_5_1 Nature of your problem D?

Explain :

____

L_5_2 What would improve the situation relating to problem D?

Explain:

____

L_6 Problem E (state code from question L_1) L_ ___________ [L_6_A] _________ [L_6_B]

L_6_1 Nature of your problem E?

Explain :

____

L_6_2 What would improve the situation relating to problem E?

Explain:

____

INSTITUTIONS SUPPORTING SOUTH AFRICAN SMALL-SCALE CONSTRUCTION ENTERPRISES

F.A   CONTRACTORS' ASSOCIATIONS

F.A.1   The African Builders Association of Southern Africa (ABA)

(a) The establishment of ABA : ABA was founded in April 1988 as the Transvaal African Builders Association (TABA) as a result of the realisation that black builders faced similar problems and shared common aspirations and goals. ABA is affiliated to the Foundation for African Business and Consumer Services (FABCOS), a development organisation promoting the economic empowerment of black businesses and consumers.

FABCOS raised the question of the general lack of self-confidence among Blacks, particularly in business. ABA's president, Mr. J. Mogale, believes that this may be attributable to Blacks having had relatively little experience in the economic arena. ABA aims to organise and develop black builders and tradesmen in order that they might begin to assume a role in the construction industry and in the development of black communities. ABA's vision is to develop sense of identity and belonging among black builders in the belief that this would increase their effectiveness and capacity (Calcopietro, 1992).

ABA has provincial branches in the Transvaal, Natal, Western Cape and Orange Free State and is reportedly negotiating the establishment of the Eastern Cape and Transkei branches. Once this has been accomplished, ABA will have a nationwide presence. Members are carefully screened in order to reduce dishonesty in the industry and to protect consumers and the name of the association. In 1990 Mogale claimed that ABA "now represents some 95% of builders countrywide" (ABC, 1990d). At the end of 1991 the organisation had 3000 members (Calcopietro, 1992).

(b) ABA's Objectives : ABA's principal objective is to facilitate the economic empowerment of black builders. The organisation seeks to accomplish this by (Calcopietro, 1992):-

(i)   Rallying African builders in order to encourage and promote affiliation to local associations for the purpose building a national body.
(ii)   Acting as a mouthpiece for members while serving as a forum within which African builders can develop solutions on matters affecting them as individuals and encouraging training and the upgrading of business skills.
(iii)   Establishing standing committees to study, research and mobilise expertise in the industry.
(iv)   Maintaining high professional standards through the establishment of sub-committees and standing committee which function as the base of knowledge and expertise in the industry.
(v)   Establishing a marketing information service for builders.
(vi)   Focusing on affirmative action and building industrial relations.
(vii)   Establishing a trust fund for the procurement of land and development finance.
(viii) Encouraging members to create new employment opportunities for black communities.

(c) Activities of ABA : The institution currently offers the following services (Calcopietro, 1992) :-

(i)   Management and other business skills training for ABA members. This programme made use of EDSA's training modules and was funded by a $100 000 grant received from the United States Agency for International Development (USAID).
(ii)   Building materials outlets in various localities. Members are invited to buy shares in and profit from these establishments.
(iii)   ABA is currently investigating the manufacture of building materials and the negotiation of franchise opportunities which members would own and operate.
(iv)   ABA MARKETING undertakes to cater for the financial, training and business opportunity needs of black builders. The organisation carries out need-analysis surveys among members in order to determine priorities and plan strategies.

(v)   In response to the general lack of skills in project financial management, workshops and training programmes are organised with the aim of upgrading members and facilitating professionalism and entrepreneurship.

F.A.2   National African Federated Chamber for the Building Industry (NAFBI)

(a) The establishment of NAFBI : NAFBI was formed in May 1991 and is affiliated to the National African Federated Chamber of Commerce & Industry (NAFCOC) (NAFBI, 1991). The establishment of NAFBI was the result of the amalgamation of two groups - the Housing Committee of NAFCOC and the Professional Builders Federation (PBF).

The PBF held its inaugural annual general meeting in November 1988 (Padi, 1988a) and occupied permanent offices in Johannesburg in January 1990. Membership included many contractors' associations and individuals countrywide and reportedly the organisation had over 2 500 applications on file in February 1990 (ABC, 1990b).

The executive committee of the PBF had succeeded in negotiating the inclusion of its members on housing projects with SAHT, HODECO, and the UF. In January 1990 its members were reported to be involved in projects worth over R25 million in the Transvaal (at Vosloorus, Daveyton, Sebokeng, Soweto, Kanana, Jouberton and Kutlwanong), the Eastern Cape (Motherwell) and the Western Cape (Khayelitsha) (ABC, 1990b).

An important contribution to the support of black SSCEs by the PBF was its official mouthpiece, the journal African Building Contractor (ABC). This publication has done much to expose the problems and successes of this group of contractors and, through its instructional articles, has contributed to their education. The publication of the journal has fortunately continued, despite the disbandment of the PBF.

(b) NAFBI's objectives : The NAFBI constitution describes the organisation's objectives as follows :

"(i)   to promote the interests of the building and allied trades and supporting services and professions in Southern Africa, in particular such of those businesses as are owner-managed;
(ii)   to promote the spread of entrepreneurship in the building industry;
(iii)   to obtain the removal of legal and other restrictions which hamper start-up and small building contracting activities or inhibit the development of activities in the building industry;
(iv)   to facilitate the provision of training, counselling and advisory services to members;
(v)   to facilitate the supply to members of purchased materials at competitive prices;
(vi)   and to communicate regularly with members, on matters of interest to them, by way *inter alia* of newsletters, meetings, seminars and conferences." (NAFBI, 1991).

Clause 23 of the NAFBI constitution provides for its affiliation to NAFCOC. This involves a written agreement between the two organisations, an affiliation fee and an obligation for NAFBI to support NAFCOC by promoting its objectives (NAFBI, 1991).

(c) NAFBI's plans for the future : A recent NAFBI document proposes the establishment of technical aid centres. NAFBI plans to address the unresolved problems which have surfaced in the programmes run by organisations like the UF, the Independent Development Trust (IDT), the SAHT, the DBSA and others. NAFBI believes that this could be accomplished via the establishment of technical aid centres in key localities across the country. The purpose of these centres would be to serve black SSCEs and prospective home-owners across a broad range of issues relating to housing and the building industry (NAFBI, 1992).

Although NAFBI's thinking has not progressed beyond the conceptual stage, it envisages that the centres would offer the following services to members :

"(i)   *Estimates and cost advice*
(ii)   *Cost planning and control*
(iii)   *Tendering and contractual arrangements* - advise and guide members on the choice of method of

appropriate tendering procedure and subsequent contractual arrangement

(iv) *Planning and programming* - advice on forward planning and programming of works to ensure efficient and timeous completion thereby maximising returns. Also guide on ordering and purchasi materials and selection of sub-contractors

(v) *Interim valuations* - measure, value and prepare recommendations for interim payment from client payments to suppliers and sub-contractors and monitoring cash flow according to cost plan

(vi) *Final accounts* - prepare, negotiate, agree and settle final accounts with clients, suppliers and contractors

(vii) *Act in dispute* - mediate, negotiate and provide expert opinion/evidence in any dispute arising from property and construction contracts" (NAFBI, 1992).

F.A.3   Other Contractors' Associations

It was evident from the literature survey that there has been a tendency for SSCEs to form associations. The princip aim of most of these associations has been to make representations, on behalf of their members, to institutions and bodies which are in a position to allocate sites or offer building work (ABC, 1990g).

It was not possible to identify all of these associations, but those that were mentioned in the literature are listed bele along with whatever detail could be gleaned.

(a) Vosloorus Builders and Allied Trade Association: The association was formed in April 1982 with the broad objective of contributing towards development of the building industry in the East Rand region. The founder membe of this association claim to be the first to have undergone training in building administration and other courses relev to building. This group received BIFSA training and graduated with BIFSA certificates. The association reports hav received assistance from HODECO (ABC, 1988h).

(b) Soweto Building Contractors' Association : This association was formed in 1986. Its constitution is aimed at: organising, encouraging and promoting the spirit of unity among builders and the public; and setting and maintainir the highest professional standards among its members. To this end it has a disciplinary committee to control misconduct, for example, dishonesty of members in their dealings with clients. The association has reportedly established good relationships with government departments, community councils, large corporations and building industry companies (ABC, 1989h).

(c) Ennerdale Builders Guild : The coloured builders of Ennerdale formed the Guild in July 1987 with the main objective of promoting and working towards free enterprise. It was intended that the Guild would represent the interests of members to relevant authorities and other institutions involved in building (ABC, 1988e).

In addition to the above, the following institutions are referred to in the literature, but no details are given of their formation and activities :

Alexandra Building Contractors' Association (ABC, 1989f)
Black Builders' Association (SAHAC, 1992)
Dobsonville Builders & Allied Trades Association (Padi, 1989a)
Daveyton Builders' Association (Padi, 1990b)
Golden Triangle Black Builders' Associations (ABC, 1989g)
International Roots Builders' Association (Padi, 1990b)
Newcastle Builders' Association (Padi, 1990a)
United Black Builders' Association of SA (HSA, 1989c)

F.B   PROFIT-MAKING DEVELOPMENT INSTITUTIONS

F.B.1   The Small Business Development Corporation Limited (SBDC)

(a) The establishment of the SBDC: The SBDC was incorporated in 1981 as a public company with an authorised capital of 150 million shares of R1 each, split equally between the public and private sectors. In terms of the Small Business Development Act, (Act No. 112 of 1981), the personnel and assets of the Development and Finance Corporation Limited (DFC) and the Indian Industrial Development Corporation Limited (IIDC) were transferred to the SBDC and these two corporations thenceforth ceased to exist (Courier, 1991:3).

(b) The SBDC's objectives: The SBDC's name implies that its prime objective is the development of small business. While this is no doubt true, it must be recognised that, as a profit-making venture, it would only do this if it were profitable. The company's objectives are concisely summarised in its mission statement:

> "...to harness the power of entrepreneurship by developing small business for the benefit of all South Africans. In striving to attain this mission, the SBDC acts in four key areas:

- The provision of finance for small business in both the formal and informal sectors.

- The provision of affordable business premises in neglected areas.

- Development support services, for example advisory support services, information and training for entrepreneurs as well as marketing support.

- Development promotion services which include community development, and various projects promoting the wider interests of the small business community." (Courier, 1991).

(c) Activities of the SBDC: The SBDC's track record over the first decade of its existence includes: the development of 800 000 square metres of affordable business space valued at a total of R280 million, throughout South Africa; the purchase and conversion of certain redundant buildings into industrial 'hives'; the award of over 30 000 loans totalling over R1,2 billion; the provision of assistance through the dissemination of information and advice to approximately one million people (at a rate of approximately 17 000 enquiries per month); and the creation and maintenance of about 285 000 employment opportunities at an average cost of less than R4 000 each (Courier, 1991:3). These are, however, general achievements which relate to all sectors of the small business environment.

The vehicle for the SBDC's interaction with small builders is the *small builders' bridging fund programme*, through which it makes loans to SSCEs. The purpose of the fund is to provide finance to SSCEs who cannot usually raise loans on the open market. Loans are currently offered in the Western Cape at 18% per annum up to a maximum of R40 000 per contract and limited to R100 000 per builder. In the Durban branch, the loan amount may not exceed one-third of the total contract sum. To qualify for a loan the applicant must; have some experience in building; must have basic building equipment; and have sufficient practical experience and management skills to successfully complete a building project. In addition, the applicant: is required to arrange for his client's financial institution to cede the contract payments to the SBDC; must produce a signed building contract, a detailed cost breakdown, a statement of projected cash flow and confirmation from the project financier that funds are available (Calcopietro, 1992). Over and above this, the contractors must waive his lien, and must preferably have a contractors' all-risks insurance policy.

F.B.2   SA Housing Trust (SAHT)

(a) The establishment of the SAHT : The SAHT incorporated as a public company in November 1986, with the South African Government providing an initial interest-free R 400 million loan. A further R 800 million was to be raised by the private sector (HSA, 1989e).

(b) The objectives of the SAHT : In 1991 the Trust's mission statement was given as follows :

> "To promote and facilitate the provision and funding of affordable shelter, essential community requireme
> and security of tenure for the lower income communities of South Africa in accordance with; the current a
> anticipated future needs, aspirations and viability of these communities by; offering appropriate services w
> will co-ordinate the efforts of public and private sector organisations and other institutions which could - a
> should - participate in the total delivery process within the economic realities of this country; in close liais
> with the communities concerned" (ABC, 1991d).

In addition, the SAHT endeavours to aid low-income communities via the provision of housing and the creation of
and in this regard, the initial focus has been on black communities. In pursuance of its goals the SAHT liaises and
cooperates with local authorities and training centres (HSA, 1989e).

(c) Structure of the SAHT : A full-time executive management team heads the SAHT. This executive's functions a

(i)   the generation of funds
(ii)   project development and assessment
(iii)   the disbursement of housing loans - this is done through the SAHT subsidiary, Khayelethu Home Loans (
        Ltd. (KHL). KHL operates through independent branch offices known as Residential Agents (RA) which a
        established in each development area
(iv)   the acquisition of land
(v)   the provision of serviced stands
(vi)   the use of bulk-buying muscle and the transference of savings to project teams
(vii)   general administration
(viii) the monitoring of projects in order to ensure an acceptable level of technical business and ethical standard
        (HSA, 1989e)

In addition, a specialist team conducts surveys, research and development, attending to the technical support of site
personnel (HSA, 1989e).

The educational function of the SAHT includes the encouragement of entrepreneurs within communities to enter th
mass housing field. Specialists are stationed on, or in the vicinity of, development sites in order to inform and assi
home builders (HSA, 1989e).

(d) The Small Builders Department (SBD) of the SAHT : This Department was established in 1989 and was
originally called the Project Management Department (PMD). Through the Small Builders Development Programm
(SBDP) of this Department and its agencies - Development Agents (DA) and Construction Management Agents (C
-, the SAHT endeavoured to maximise the involvement of SSCEs in its projects.

The DAs were established housing developers who built and sold houses for the SAHT, while CMAs administered
management and logistical support to small local entrepreneurs on SAHT projects. The CMAs strove to achieve th
by assisting with: project and construction planning; administration and reporting; purchasing of materials; routing
control of materials; supervision of quality; training and skills transfer; and marketing.

The CMAs worked in conjunction with the SAHT's PMD. This Department organised the manner in which the CI
interfaced with the SSCEs and ensured the effective transfer of skills. A further function of the Department entaile
the provision of serviced land, bulk-buying facilities and contract documentation (including drawings, the specifica
and bills of quantities) (HSA, 1989e).

The activities of the SBD attracted criticism from builders working under the CMA system and "influential black
groupings and chambers of commerce" (Duncan, 1991). This criticism was mainly directed at the CMA concept a
the development programmes. In response to this, the SBD consulted with black builder and commercial groups
(Duncan, 1992a), consequently revised its thinking on SSCEs in 1991, and adopted a new methodology. Duncan
(1992b) described its underlying principles as follows:

> "Only business and management skills would be taught, and technical competency would be a prerequisite

gain entry to the programme.

The road to self reliance is a process and therefore the development of the builder must be continuous, i.e. from Small Builder to Mini Developer to Developing Agent. The fundamental differences between the three being the amount and type of assistance offered by the Department and conversely, the amount of risks and responsibility taken by the builder.

... strict entry criteria are enforced to safeguard ourselves and to ensure that the builder has a better than even chance of being successful.

A strong emphasis is made on utilising local builders and labourers for the purpose of retaining the newly acquired skills and the wealth created in the community, once the SAHT has completed the project and left the area.

A structured theoretical curriculum with built in tests will be presented by external professional trainers with a minimum pass rate required to remain on the programme".

In this new initiative the SBD effectively operates as an autonomous development company within the SAHT and currently accounts for 20% of total SAHT production. The Department's mission is twofold - to uplift builders and to provide affordable homes (Duncan, 1992b).

Under the new development programme, participating builders are evaluated against a set of entry criteria. The entry criteria for Small Builders and Mini Developers are as follows:

Small Builder's Qualifying Criteria (Duncan, 1991): In order to qualify, applicants must;

(i)   be literate, numerate and capable of reading drawings; be able to set out foundations; be able to cost labour, materials and own overheads; be able to compile material requisitions; and must have an elementary understanding of legal contracts.
(ii)   have access to an office.
(iii)   keep records of labour and material costing and business as a whole;
(iv)   be sufficiently well resourced to be capable of simultaneously constructing five 42 m^2 houses;
(v)   operate a bank account;
(vi)   show proof of access to bridging finance sufficient for one month;
(vii)   employ own contracting staff or sub-contractors (bricklayers, plasterers, carpenters, roofers and painters);
(viii)   have access to specialist contractors - glazing, ceiling, plumbing and electrical;
(ix)   be prepared to personally supervise site operations on a full-time basis;
(x)   have access to plant - scaffolding and trestles, hand tools, light delivery vehicle, compactor and power floating equipment;
(xi)   produce written references from clients demonstrating a three year track record of acceptable standard and quality of work;
(xii)   Must be capable of completing five 42 m^2 houses in a six week period.

Mini Developer's Qualifying Criteria (Duncan, 1991): Applicants must;

(i)   be sufficiently well resourced to simultaneously construct at least eleven houses;
(ii)   produce a R5 000 cash deposit to be paid into a trust account, or a bank guarantee;
(iii)   have at least a "C" credit rating;
(iv)   have personnel and equipment as detailed above for Small Builders (see items (vii), (viii) and (x));
(v)   produce proof of a significant track record in the home-building industry, supported by at least three references from previous employers or clients;
(vi)   produce proof of acceptability of quality of previous work - this must include ten affidavits from home owners and two recommendations from quality inspectors at financial institutions;
(vii)   produce proof of an acceptable materials purchasing record from reputable suppliers, over the period of one year;
(viii)   produce evidence of; experience and skill in marketing of houses; procurement, distribution and consumption

of materials; and business administration, financial and accounting systems;

(ix)   produce a detailed organisation chart detailing firm's structure and staffing;

(x)   employ an experienced and capable full-time site supervisor, suitably qualified in technical and communicat
skills.

The new Development Programme : Having met these entry criteria, participant builders and developers are evaluat
by a SBD tutor and areas of weakness are identified. Duncan (1992a) explains the training programme as follows :

> "The Competency Based Modular Training (CBMT) will be informal and individually targeted (1:1) where
> builders will learn at their own pace from a lecturer on site. They will only be taught what they lack of the
> specific criteria and knowledge necessary to fulfil their functions. The rationale being that not all builders
> possess the same skills and any training that is not specific, is effectively wasted.
>
> The course will be divided into two phases with eight sections which will be divided into a further fifty
> separate modules and designed so that each module should be completed in one training session with one pe
> week. At the end of this period, a period of a six weeks (one full day per week) will be spent with each
> builder ensuring that everything he has learned is put into practise. At the end of this six week period a
> diploma will be awarded to those builders having been successful at the 'theory' and the 'practical'."

Phase 1 of the course covers contract management (tendering for, planning and executing the contract), while Phase
focuses on managing a homebuilding firm (marketing, finance, administration, manpower and establishing the
business) (Duncan, 1992a). Table FB.1 summarises the support offered by the SAHT to the various types of
participating contractors.

	TYPE OF SUPPORT	BENEFICIARY		
		SMALL BUILDER	MINI DEVELOPER	DEVELOPING AGENT
1	Bridging Finance (short term loan)		✓	
2	Working Capital (payment of creditors)	✓		
3	Bond Finance (subject to availability)	✓	✓	✓
4	Serviced Stands (on SAHT projects)	✓	✓	✓
5	Formal Instruction (lectures and CBMT)	✓	✓	
6	Supervision (constant guidance by site manager)	✓		
7	Marketing and Sales (totally or partly)	✓	✓	
8	Continuity of work (as far as possible, but limited to 2 years)	✓	✓	
9	Individual Counselling (full-time dedicated manager)	✓	✓	
10	Access to specialists (engineers, quantity surveyors, etc.)	✓	✓	
11	Estimating and Pricing (labour rates and quantities)	✓		

Source : Duncan (1992a: 7)

Table FB.1 - Support Offered by the Small Builders Department of the SAHT

(e) SAHT's achievements : The following brief summary of selected SAHT achievements is included in order to enable the reader to assess the effectiveness of the SAHT and the extent to which SSCEs have obtained access to work through its developments.

Under the old SBDP, 850 stands in Alexandra were allocated for financing by SAHT, and development by Murray & Roberts, its partner in the scheme. Of these stands, 700 went to Conde (North) (Pty.) Ltd., with Ribco as approved sub-contractors. 150 Low-cost and 550 high-cost houses were erected (ABC, 1989e).

The SAHT retained the remaining 150 stands which were allocated as follows: 69 to Chandor training centre (which trains black builders); 54 to Ivory Tusk Construction - an independent entrepreneur (who worked under the supervision of a CMA appointed by the SAHT); 9 to the Alexandra Building Contractors' Association (for distribution amongst its members); 3 were reserved for experimental housing; and 15 were allocated to owner-builders (ABC, 1989e).

The development at Jouberton was another SAHT project executed under the old SBDP with 150 stands being allocated to black builders. These builders were introduced to the CMA (Civimech Project Management Services) by the Golden Triangle Black Builders' Association. During this project, the CMA discovered that none of the black builders employed on the project could complete a house from start to finish. However, these have reportedly all gone on to become SSCEs running their own banking accounts (ABC, 1989g).

The Palm Springs development near Soweto, will involve the SAHT in the development of 22 000 residential stands. Work is expected to continue until 1996-1997, and this should provide a useful testing ground for the SBD's new development programme (ABC, 1990e).

During 1989-1990, 359 artisans and 1231 labourers participated in the old SBDP and 121 SSCEs emerged as developers in their own right (ABC, 1990f; ABC, 1991c). In 1991 it was reported that, since the inception of the SBDP, 242 local builders had been assisted in gaining entry into the construction market, constructing some 2665 units (ABC, 1991d). Notably this figure indicates that black SSCEs had only been responsible for about 9 % of all houses developed by SAHT - (in September 1991 it was reported that, since 1987, the SAHT had facilitated the building of 28 345 homes (ABC, 1991d)).

In 1992 Kevin Duncan, head of the SBD, gave a revised report of these data, indicating that

> "10 Mini-Developers, 112 Small Builders, 458 artisans and approximately 2000 labourers have benefitted from participation in the programme. 3 508 Houses have been constructed with a total contract value of approximately R 70 million, affording shelter to nearly 17 000 people. Currently on SAHT projects throughout South Africa, 8 Mini-Developers, 31 Small Builders, 206 artisans and approximately 700 labourers have work in progress of approximately R7,5 million and are enjoying meaningful employment under the SAHT Small Builder Programme" (Duncan, 1992a).

In the latter part of 1991, it was reported that the Board of the SAHT had approved future projects totalling R1,4 billion which will result in the erection of 67000 houses and the servicing of 68000 stands (ABC, 1991d).

F.B.3   The Kwazulu Finance and Investment Corporation Limited (KFC)

The region of Kwazulu is similar to the Transkei, Boputhatswana, Venda and Ciskei, in that it has its own government, but unlike the latter four, it has never become independent. As part of its Apartheid plan the South African government tried to stimulate development within the territories that it regarded as black homelands (mainly rural areas) in an effort to transform them into self-sufficient mini-countries. The vehicle it invented for this process of development was the statutory development corporations. Because the scope of this dissertation is limited to exclude the rural areas, the activities of the numerous development corporations will not be reviewed. But there is one important exception - the case of Kwazulu. The degree to which the metropolitan regions of Durban-Pietermaritzburg intermesh with the territory of Kwazulu makes it impossible to think of them as separate territories. Parts of Kwazulu are indeed as urban as Durban, while other parts are definitely rural. The inclusion of KFC in this section is thus

necessary because its activities are very much a part of the lives of Durban's small contractors. A similar case could
of course be argued for Boputhatswana and its proximity to Johannesburg, but since it is not considered to be a part
South Africa, its development corporation has been excluded from this section.

It may seem inappropriate to categorise an institution such as KFC as profit-making. However, since it has issued
ordinary and preference shares and since it pays dividends to the latter (KFC, 1992), it clearly must be seen as a
profit-making development institution (PMDI).

(a) The establishment of KFC: KFC is a statutory development corporation which was established in 1978 by
proclamation R73, in terms of Act No. 46 of 1968, as amended by the Kwazulu Corporations Act (Act No. 14 of
1884). The Kwazulu government holds all of KDC's ordinary shares and its 10-member board reports directly to th
Kwazulu Legislative Assembly.

(b) KFC's objectives: The mission of KFC states that it aims "to significantly contribute to the socio-economic
empowerment of the people of Kwazulu/Natal" (KFC, 1992). This particular focus on Kwazulu/Natal must be seen
perspective - the region occupies approximately 9% of the Republic of South Africa, and with a population of 8,7
million, it represents 22% of the country's population. Unemployment is currently about 20%, while the illiteracy r
as measured in the 1985 population census was 70%. The objectives of the organisation are to (KFC, 1992):

(i)   stimulate, promote and sustain entrepreneurship in all sectors in [Kwazulu];"
(ii)   contribute towards income generation within [Kwazulu];"
(iii)   contribute towards the generation of wealth and its distribution to the target population;"
(iv)   stimulate and contribute to the provision of accommodation in [Kwazulu], especially home-ownership, in
         order to enhance the quality of life, as well as to provide opportunities for private sector investment.

(c) KFC's achievements: KFC's main activities in pursuance of these objectives are (KFC, 1992):

(i)   financial advances and investments;
(ii)   the creation of employment and income-earning opportunities;
(iii)   entrepreneurial development in the commercial, industrial, agricultural and mining sectors;
(iv)   the provision of appropriate physical and business infrastructure;
(v)   industrial recruitment under the Regional Industrial Development Programme;
(vi)   the establishment of major commercial undertakings and facilities;
(vii)   the provision of training and consultancy services (in association with Kwazulu Training Trust);
(viii)   the provision of housing and the development of residential infrastructure.

Since 1979 KFC has achieved the following: a cumulative investment in Kwazulu of R1,167 billion; the advanceme
of loans totalling R620,7 million; an investment in land and buildings of R550,5 million; the creation of 80 000 job
the establishment of 6 000 enterprises; and the training of 30 000 people (KFC, 1992). The corporation's specific
involvement in housing can be gleaned from its resume' of "recent" major housing projects in its 1992 annual repor
as follows (KFC, 1992): sites developed and serviced - 2127; homes constructed and financed - 1219; and homes
financed - 839.

F.B.4   HODECO (Pty.) Ltd. (HODECO)

(a) The rise and fall of HODECO : The organisation was established in mid-1986 by the Perm as an ordinary
subsidiary company. In essence, HODECO sought to address the problems of black SSCEs in the homebuilding
market. The Perm maintained that black builders should have been involved in attempts to solve South Africa's
housing crisis and deserved opportunities to develop their skills (Creighton, 1992). Although HODECO was liquida
late in 1992, its objectives and achievements are reviewed below because they clearly reveal something that did not
work.

(b) Objectives of HODECO : Creighton (1992) expressed HODECO's mission as follows :

> "In order to create an affirmative action programme to address the difficulties of SSCEs, a Home Development Company (HODECO) was launched by the Perm ... as a social responsibility operation to facilitate the entrepreneurial development of the black building contractor by assisting him to get a share of the action by building affordable housing for HODECO to market on his behalf."

(c) HODECO's involvement with black SSCEs: HODECO's analysis of the problems experienced by black SSCEs revealed that: the lack of access to land; the inability to package a viable contract; and lack of access to bridging finance presented serious barriers for black SSCEs wishing to enter the homebuilding market (Creighton, 1992).

For the first three years of its operations, HODECO procured rights to serviced sites, packaged a range of affordable homes, and contracted with black SSCEs to construct show houses. These houses were then sold with the help of black estate agents. However, the increasing level of violence, intimidation and vandalism in the black townships caused HODECO to reassess its position and triggered a change in policy (Creighton, 1992).

This policy change involved a greater focus on the need for the black SSCE to become more orientated towards marketing himself and his product. In practice, HODECO contracted with the SSCE for the full value of labour and materials and also provided guidance and financial assistance. In this manner HODECO effectively employed the small contractors to build speculative housing. An added advantage to the SSCE was that, through the Perm, HODECO was able to provide mortgage bonds to the buyers of houses produced by the SSCEs. HODECO strongly assisted its builders with promotion and selling, particularly into the corporate employer market. A further role played by HODECO involved introducing builders to other network links, for example, bridging finance through the SBDC or EDSA (Creighton, 1992).

(d) HODECO's achievements : In February 1992, Creighton (1992) reported that HODECO had completed roughly 650 houses in the past 4½ years. These were mainly developments in Soweto, East Rand and Vaal Triangle.

HODECO has worked with 192 SSCEs since inception. Most of these builders only built 1 house (32%), while another large group (23%) built 2 houses. Forty-one percent (41%) of the SSCEs produced more than 2, but fewer than 10 houses and the vast minority (4%) built more than 9, but fewer than 22 houses.

(e) Nature of assistance under the HODECO operation : Creighton (1992) provides the following details of the HODECO operation :

Normally the builder identifies the customer, while under HODECO the builder was assisted in this, sometimes to the extent of full marketing. Costing and contractual functions are usually the builder's responsibility, but HODECO assisted its builders in these. HODECO provided its builders with approved house plans, whereas the builder would usually produce these using his own draughtsperson.

Finance was relatively guaranteed for qualified buyers assisted through HODECO, whereas usually the builder and customer would have to arrange this. Further, access to land was facilitated by HODECO and builders were provided with serviced and paid-for sites. Access to bridging finance through EDSA or the SBDC would usually be arranged by the builder in co-operation with a financier, while finance from these sources was relatively easier to obtain for HODECO builders, since HODECO ceded the contract price.

F.C.1   The Urban Foundation (UF)

(a) The establishment of the UF : The UF was established in 1977 as a non-profit-making corporation by leaders
the South African business community with the support of black community leaders (UF, 1989a). The organisatio
an independent development group, accountable to its board of governors, operating from six local and two
international offices. These are situated at Johannesburg (national office), Bloemfontein, Cape Town, Durban, Por
Elizabeth, London and New York (UF, 1990).

(b) The objectives of the UF : In 1992 the UF mission statement read as follows :

> "[The Urban Foundation's] mission is to promote development that empowers people and communities to
> establish control over their lives. Its particular focus is on finding sustainable solutions to the critical prob
> facing poor people, especially those in and around the cities. It carries out its mission by; formulating and
> promoting the adoption and implementation of effective development policies; initiating and establishing s
> reliant institutions with the capacity to address critical needs and that share The Urban Foundation's value
> and approach to development; and supporting communities and organisations involved in development" (U
> 1992).

(c) The Housing Utility Companies (HUC) of the UF : Prior to 1991, the UF consisted of two main divisions, n
the Residential Development Division (RDD) and the Policy and Programmes Division (UF, 1990).

The old RDD was made up of a housing policy research unit and the HUCs. These utility companies were all
incorporated as Section 21 (non-profit) companies. The RDD received an initial grant of R4,25 million from the U
but has subsequently conducted its operations on a commercial basis, borrowing from the UF reserve fund and oth
financial institutions. All such loans have been procured and controlled centrally by the UF (UF, 1989b).

The aim of the RDD was: to play an effective role in the provision of serviced sites and affordable housing (this
included the provision of affordable housing, self-help schemes and informal settlement upgrading); to perform thi
role in close collaboration with the affected communities; to facilitate the involvement of the homebuilding sector
as employees and as self-employed contractors and sub-contractors; to conduct research aimed at improving
affordability and to disseminate research findings to all other participants in the provision of housing; and to resea
all financial and socio-economic factors which obstructed these objectives with the purpose of promoting policies a
strategies to overcome these (UF, 1988).

For the sake of clarity, the history and activities of the HUCs are now briefly summarised. The first HUC was
established in 1983 and in 1988 there were seven of these. Their principal activity was the development of service
residential land and homes (in the R5 000 to R35 000 category) for lower and middle income urban communities.
Each company was controlled by a board of directors which included local business and community leaders and U
representatives (UF, 1988; UF, 1989b).

In 1988 the seven HUCs (*Azalea* in Pietermaritzburg, *Blomanda Housing Company* in Bloemfontein, *Family Hous
Association* (FHA) in Johannesburg, *Innova* in Durban, *Masizakhe Utility Company* in East London, *Unifound* Po
Elizabeth and *Cape Utility Homes* in Cape Town) generated site sales and housing worth R123 million (UF, 1988
by 1989 a total of 26 564 serviced stands and 18 059 houses had been provided (UF, 1989a). In 1990 the HUCs
experienced a 26% reduction in real terms on the previous year's operations and collectively lost R1,9 million. Th
loss was attributable to inflation, high interest rates and high levels of violence in development areas (UF, 1990).
Matters deteriorated in 1991 with the utility companies incurring an operating loss of R 17 million on a turnover c
101,7 million. This loss mostly represented "interest incurred on investment in serviced stands and the cost of
rectifying defects in houses sold several years ago in the Eastern Cape and Natal" (UF, 1991).

The changing political scenario, the ongoing high level of violence, and high interest rates prompted the UF to

radically restructure the RDD. This restructuring (funded by a R 15 million grant from UF reserves, which "...
included an amount of R7,9 million in respect of losses incurred in treasury activities undertaken to supplement the
financing of the housing programme" (UF, 1991)) involved: the merger of the four remaining HUCs into one
company, Innova and the establishment of the Land Investment Trust (LIT) - which obtained its initial capital of R70
million from the IDT. Innova and the LIT together form The New Housing Company Group (The Newco Group)
which is legally and financially independent of the UF. The LIT was established to provide the funding for the land
related activities of The Newco Group and to provide loans for land development to other agencies involved in low-
income shelter projects. Four of the old HUCs (*Blomanda Housing Company, FHA Homes, Innova (Natal)* and
Unifound) are now the operating divisions of Innova. The focus of The Newco Group's activities is on developing sites
under the IDT's capital subsidy scheme and acting as project managers in the upgrading of informal settlements (UF,
1991).

A current informal settlement upgrading project is Besters Camp (Durban) where 8 250 structures are being built for
some 57 000 people. In this project all work is being carried out by contractors drawn from the community and who
are trained and supervised by the Natal region's informal settlements division. Plans for similar projects in the future
include: Soweto-on-Sea (Port Elizabeth), an upgrading scheme involving 90 000 inhabitants; and Freedom Square
(Bloemfontein) involving 4 000 households (UF, 1991).

(d) The Community Enablement Programme (CEP) of the UF : "The [CEP] is designed to support community
initiatives to bring about positive and substantial change in people's lives. This is done through specially designed
resource centres developed in local communities, and through initiatives to promote formal and informal economic
activity. During 1990 the Foundation expanded its resource centre programme (see Urban Foundation Annual Reviews
for details concerning the locations and activities of the resource centres) throughout South Africa and began to
establish its three economic advancement projects as independent ventures" (UF, 1991). These projects are: the Private
Sector Counselling Organisation (PRISCO); the Contractor Development Agency (CDA); and the Sunnyside Group
(UF, 1991). These projects are briefly described below.

PRISCO : This programme is directed at improving the management competency of black entrepreneur-owned
businesses in the informal sector. Since its inception in 1987, PRISCO has counselled 425 businesses and currently
operates in twelve PWV towns (UF, 1991).

CDA : The focus of the CEP's promotion of formal sector economic activity is on the black building industry. This
led to the establishment, in 1988, of the CDA which sought to address the concern among black building contractors
that they were unable to compete against established corporate developers in the black homebuilding market. The CDA
offers to maximise the management skills of leading black contractors and to ensure that they have access to sufficient
loan capital (UF, 1989b).

The CDA project was piloted in 1990-1991 in the PWV area, attracting twenty participants by mid 1991 (UF, 1990;
UF, 1991). The project relies on the cooperation of three banks (African Bank, Standard Bank and Trust Bank) which
provide the financing. Credit is 60% guaranteed by the DBSA, while access to serviced sites is facilitated through
FHA, SAHT and other developers (UF, 1990).

The programme is specifically targeted at "A-Grade" contractors. To qualify, the contractor must have a minimum
annual turnover of R 300 000 and a history of satisfied clients. In addition, he must have a sound financial history.
Prior to his being accepted on the programme, the technical and management competence of each contractor is
assessed and a bank of his choice is approached as the proposed financier. This bank then scrutinises the contractor's
banking record and if this is found to be in order, the CDA offers to sign a contract with the builder (ABC, 1991a).

The contract offers to: pursue overdraft facilities for individual contractors; to tutor contractors or their delegates in
the operation of the core management, financial and planning modules; and to ensure a steady inflow of work for each
contractor (ABC, 1991a). The contractors are required to pay a levy based on turnover (3½ % of all work generated
through the activities and financing of the CDA) and an initial entry fee. The R5 000 entry fee entitles the contractor
to use all of the programme's computer software packages, which may only be employed in his business once the
CDA management team is satisfied of his competence. Plans for the future include extending the CDA programme to
reach less advanced labour-only and labour-and-material contractors (ABC, 1991a).

Building contracts worth R 3,5 million were completed by CDA participants during 1990 and further contracts to the value of R 50 million were under negotiation in 1991 (UF, 1991).

The Sunnyside Group : This group was formed in 1987 as a representative body for small enterprise in South Africa and is supported by the UF. The group is a lobbying organisation made up of fifty development agencies. Its main function is to promote appropriate regulation (UF, 1990).

F.C.2   Entrepreneurial Development (Southern Africa) (EDSA)

(a) The establishment of EDSA : The organisation was established in September 1990 (EDSA, 1990) and was incorporated under Section 21 of the Companies Act, 1973, not for gain. EDSA is financed by grants and sponsors of a general nature or for specific projects (ABC, 1990c). The organisation came into being as the result of a merger of two existing organisations, namely the IBM Small Builders Bridging Finance Loan Scheme and the Foundation for Entrepreneurial Skills Training (FEST) (ABC, 1990c). The FEST programme had initially been established as a BIFSA project.

(b) Objective of EDSA : EDSA was formed to "develop the rapidly growing number of small building contractors operating predominantly in the informal sector of the economy" (ABC, 1990c).

(c) Activities and achievements of EDSA : Perhaps EDSA's most important activity is its *training programme* which is intended for SSCEs throughout South Africa, free of charge. The training is designed to be relevant to the needs of SSCEs and is presented as a four-tiered programme (which conforms with the DBSA "Model for the Development of Small-scale Builders" presented in Appendix D).

The first level encompasses functional literacy and numeracy. A second (introductory) level concerns the compilation of quotations for "piece jobs". This is followed by a third (intermediate) level aimed at teaching management skills at the level of house building contracts. The fourth (advanced) level focuses on management skills at the level of multi-unit developments (ABC, 1990c).

By the end of 1992 EDSA had produced 701 introductory, and 229 intermediate graduates. Future plans for this training include extending its accessibility to neighbouring countries via franchising (EDSA, 1990; EDSA, 1991). In 1992 the *Small Civil Contractors' Entrepreneurial Course* and the *Building Materials Manufacturers' Entrepreneurial Course* were added to the training portfolio. The former is aimed at black SSCEs doing water reticulation work in Soweto, while the latter emerged after the need to develop viable building materials manufacturers was identified by both EDSA and the DBSA (EDSA, 1991).

Training materials and programme support are reportedly offered at no charge to regional development agencies and corporations. It is hoped that this will result in these institutions managing the programme on a franchise basis, and thereby making the training available to SSCEs nationwide (ABC, 1990c).

A further important activity of EDSA is the *Small Contractors Action Forum* (SCAF), formerly the Small Builders Action Forum. The SCAF is intended to be a facilitatory arena for the meeting of representatives from informal and formal sector organisations on the problems of the development of SSCEs. The forum operates on an open agenda and, indeed, encourages all participation (ABC, 1990c).

The *Bridging Finance Loan Scheme* is another important EDSA service to SSCEs. The Scheme began in 1986 when the African Bank signed an agreement to administer a R 300 000 fund aimed at providing bridging finance to SSCEs in Soweto. FNB later signed such an agreement and by 1989 the fund had grown to R2,3 million, and reached the whole PWV area. The fund was strengthened in 1992 by the grant of R1,7 million for 5 years from IBM (EDSA, 1992).

Under this scheme, contractors have access to bridging finance to cover the time period between building contract progress payments. Although most of the Scheme's activities have been on the Reef (ABC, 1990c), EDSA intends to expand to other parts of the country. By the end of 1990 the Scheme had made 607 loans - these resulted in housing worth R14,5 million and 10 455 jobs (EDSA, 1990). During 1992, EDSA's loan scheme assumed a new function by providing guarantees for bridging finance loans from the LIT to two black contractors (EDSA, 1992).

An important activity of the SCAF during 1991 was the production of two simplified forms of contract: an agreement for labour-only or labour-and-material sub-contracts; and an agreement for contracts involving new houses or alterations and additions to existing houses (EDSA, 1991).

In 1991 EDSA strengthened its links with the UF's Sunnyside Group and its Construction Industry Focus Team (EDSA, 1991).

F.C.3   The Kwazulu Training Trust (KTT)

(a) The establishment of KTT : KTT was incorporated in 1980 as a Section 21 company and has grown from a small training institution into a development organisation with operational expenditure of over R13 million per annum (KTT, 1992).

(b) KTT's objectives : The KTT mission is:

> "To add value to Human Resources of the region, through training and development, to enable them to become socio-economically self-reliant in a meaningful and sustainable manner" (KTT, 1992)

KTT's objectives are reflected in its programmes. Those that could directly or indirectly be aimed at SSCEs are summarised as follows (KTT, 1992):

(i)   *Community Empowerment Programme (CEP)*. This programme aims to facilitate self-reliance and self-upliftment of communities and its immediate objective is to assist communities and to establish community learning centres.

(ii)   *Programme to Improve Africa's Competence and Expertise (PACE)*. The aim of this programme is to provide technical skills to adults to improve their prospects of sustainable self-reliance. The current objective is to train 9000 people in two and a half years, in a variety of technical fields, including building.

(iii)   *Affirmative Action Programme (AAP)*. The specific aim of this programme is to facilitate the advancement of Blacks in fields where prejudice impairs their progress. The objective of this programme is to train 100 black artisans in two and a half years.

(iv)   *Entrepreneurship Development Programme (EDP)*: In this case, the programme aims to accelerate the supply of trained entrepreneurs capable of running sustainable businesses.

(c) Activities and achievements of KTT : The main activities of KTT occur under the programmes outlined above. KTT's interface with SSCEs is mainly in the PACE programme, where building skills are taught using BIFSA's training methodology. The old BIFSA Mariannhill Training College was bought for use as KTT's new training centre in 1992, funded by grants and loans from KFC, IDT and DBSA (KTT, 1992).

KTT's programme offers apprentice training in the carpentry, joinery, bricklaying, plastering, tiling, plumbing and shopfitting trades and is accredited by the Building Industry Training Board. Courses are presented on a modular basis and can be tailored to specific needs. Depending on the needs of the student or the company undergoing training, these courses are presented at basic/semi-skilled or advanced levels (Calcopietro, 1992).

In addition to skills training in the conventional trades, KTT offers training in specified skills. Further, building management courses are offered which include tuition on: tendering; planning and execution of the contract; and supervision (Calcopietro, 1992).

In addition to training, KTT provides loans to SSCEs.

F.C.4   The Palabora Foundation (PF)

(a) The establishment and objectives of the PF : The PF was established with the support of the Palabora Mining Company and is committed to the upliftment, development and advancement of people throughout South Africa. The Foundation adopts a collaborative approach, establishing long-term working relationships with the industrial informal sector, local communities and other project partners (ABC, 1991e).

The PF has established the Reef Training Centre (RTC) in pursuance of the following objectives :

"(i)   To develop a technical training centre to serve the rapidly emerging informal industry sector, commencing with the building and motor repair trades and expanding to encompass others in future;

(ii)   To create an environment which will support the development of people physically, mentally, spiritually and educationally;

(iii)   To restore the natural environment, beautify the land and create an estate that will harmonise the various training, educational and farming activities;

(iv)   To establish a farming enterprise consisting of a game farm, hay production unit, vegetable production, fish farming and a possible intensive farming endeavour. The purpose of this is to produce an income for the partial support of the training activities, and to provide a base for agricultural training and wildlife and nature conservation courses;

(v)   To establish a closer relationship with the local farming community and to serve their social and educational needs" (ABC, 1991e).

The training centre is situated on a 500 hectare farm in the PWV area. Ten hectares of this land is allocated to the building trades training programme.

(b) The RTC's building training programme : The RTC currently offers building industry skills training courses in the following trades - painting, glazing, electrical, plastering, tiling, paving, brick and blocklaying, carpentry and plumbing. According to General Manager Mr. F.J. Swanepoel (in a brief telephonic interview with the author, the training modules vary between nine and twenty-four weeks in duration and are "fully operational". This means that trainees actually build complete houses during the programme. These structures are built on the RTC's property and most are demolished on completion of the course. This is possible because a very weak mortar is used for the brickwork and most of the materials are therefore reusable. Further, trainees are taught to build the same house design with three different types of bricks/blocks. The RTC's training methodology is largely modelled on the BIFSA competency based modular training and is currently directed by a former BIFSA training manager.

(c) The RTC's plans for the future : The RTC plans to implement its Building Entrepreneurial Training Programme in mid 1992. A detailed curriculum was not yet available at the time of writing, but Mr. Swanepoel explained that it would address the common problem areas of costing, tendering, management, *etc.*. It is envisaged that this programme will be coordinated with the skills training programme in such a way that practical experience will be obtained.

F.C.5   Tholiwe Homes (TH) (TH, 1992)

(a) Objectives and strategies of TH: Tholiwe Homes was registered in 1989 as a Section 21 company, not for gain and commenced operations in 1990. The chief objectives of the company are the provision of low-cost housing and development of SSCEs originating from local communities. The strategies adopted to meet this objective are as follows:

(i)   TH uses independent builders drawn from the local community and organises them on a "voluntary group" franchise basis
(ii)   Materials are supplied, and services and training are provided with the goal of facilitating entrepreneurial development
(iii)   Normal bond finance and supplier credit are used to finance operations
(iv)   Elimination of the sub-contractor from the building process
(v)   To confine operations to the low-end of the housing market

(b) Activities and achievements of TH: TH reports that its projects have met with support from communities for the following reasons, among others: local contractors and labourers are employed; and access to facilities such as land, credit and management is facilitated. TH presently has 16 builders whom it services and who together built 100 houses in 1991.

In the proposal for a particular project, it was foreseen that 50 builders would eventually operate under TH. Training is regarded as an integral part of any TH project. This is provided, after assessing the training needs of individuals, by internal courses, adaption and presentation of external courses, or by sending the builders to external institutions. The costs of this training are built into the budget for each individual project. The types of training offered include: technical training (building skills only); project management training; and entrepreneurial training. Training is supported by counselling in the field. For this purpose, a team of counsellors (one counsellor to every ten builders) make regular visits to builders and keep records of their development.

(c) Bridging finance and supply of materials: Every builder operating under TH is entitled to make use of a bridging finance facility set up and controlled by TH. Builders pay for their own materials, but TH procures materials and later debits individual builders.

F.C.6   The Bloemfontein Building Industry Development Foundation (BBIDF)

The BBIDF is a university-based course. Its theoretical curriculum was complied under the auspices of, and is taught by, the Department of Quantity Surveying and Building Management at the University of the Orange Free State.

(a) The BBIDF Building Contractors Development Programme - Course for Small Contractors (CSC) : To gain entry to the course, contractors must have shown reasonable experience as entrepreneurs in the building industry (BBIDF, 1990). This course was first offered in 1988. Courses were presented in 1988 and 1989, but not in 1990 (due to unrest and strikes and a decline in the number of building projects available (BBIDF, 1990).

The course structure encompasses two levels - Level 1 and Level 2 - and certificates are issued on successful completion. Level 1 was presented to 4 contractors in 1989 (BBIDF, 1989) and to 14 contractors in 1991. This module runs over a four month period and consists of sixty-four half-hour contact periods (BBIDF, 1991).

Level 2 was presented to 7 contractors for the first time in 1991. To date, all Level 1 and Level 2 candidates have graduated (BBIDF, 1991).

(b) Objectives of the CSC : The objectives of the course are:

(i)   to contribute to the broad management development of smaller building contractors;
(ii)   to communicate knowledge of building technology, building industry, quantities, building finance, estimates, *etc.* to contractors attending courses - in such a manner that they can apply the knowledge practically in their businesses and in so doing, improve their businesses and;
(iii)   to acquaint student contractors with :

      *   practice procedures within the building industry
      *   technological and management principles relating to the management of small businesses

- the interpretation of architects and engineers drawings
- the use and implementation of bills of quantities as tender documents and as contract documents
- limited statutory regulations directly affecting the management of a SSCE
- the practical application of theoretical principles - this is achieved by requiring contractors to exe
  building projects (facilitated through the local Master Builders Association), under supervision, a
  practical component of the course.

F.C.7   The Shelter and Urban Development Support Programme (SUDS) (CUSSP, 1993)

This programme has not yet been implemented, but its attributes are, nevertheless, worth summarising.

SUDS is a programme which is part of a USAID initiative to:

(i) develop new ways of financing housing;
(ii) *support the development of small black contractors;*
(iii) train in the field of local government;
(iv) give legal aid to people in the field of housing, *etc.*

SUDS consists of four initiatives. The first of these is the management-oriented South African Black Contractors Assistance Programme (SABCAP). Under this project, which was initiated jointly by ABA and NAFBI, black SSC will be supported in the acquisition of business management skills (for example, estimating cost, planning material purchases, labour management, cash management, *etc.*). The teachers/counsellors on the programme will be experienced building contractors and the need for on-the-job training has been taken into account. The SABCAP programme will run for three years only and its target group will be black SSCEs who have growth potential and resources to operate a formal contract.

The second initiative is the provision of bridging finance to black SSCEs. This will take the form of amplifying th current EDSA programme by way of a USAID grant, which will allow it to expand beyond the Johannesburg area Cape Town and Durban. SSCEs will have to provide the following documentation to qualify for a loan:

(i) a formal building contract, copies of approved plans and schedules of finishes
(ii) cost estimates and details of sub-contractors
(iii) a cession whereby the SSCE waives its right to direct payment and permits an approved commer bank to pay its accounts

Firms would be limited to one loan of not exceeding R20 000 at a time.

The third initiative is a response to the building materials supply/cost problem and aims to promote timber construction, while the fourth initiative promotes the manufacture of cement-based building materials.

F.C.8   The Development Bank of Southern Africa (DBSA) and SSCEs

This section would be incomplete without a mention of the DBSA. It is really more a support environment for all the development organisations rather than for the individuals that these organisations assist. Since it does not inter directly with SSCEs, but instead does this through the medium of existing organisations, it has specifically been excluded from the categories reviewed above.

(a) The role of the DBSA in community and economic upliftment : The DBSA participates in most standing development forums, contributing knowledgable members of its staff or appointing private sector consultants. The Bank has been instrumental in promoting and facilitating interaction between public and private sector interest grou as well as between private sector sectoral groupings (HSA, 1991e).

The most important role the DBSA plays is that of funding. General Manager Johan Kruger further explains that :

> "As a 'Bank of Last Resort' [the DBSA] provides concessionary funding in most cases where it provides support, be it through technical assistance and preparations assistance loans, or in investment loans for physical development projects.
>
> It never implements projects itself, but always works through borrowers or implementing agents for borrowers or beneficiaries.
>
> The DBSA does not lend to individuals or for normal commercial activity. ... It also does not invest in land, commercial property or stock. It addresses the development needs of the poor and is a catalyst for, and facilitates development in, underdeveloped communities." (HSA, 1991e).

Examples of the type of support provided by the DBSA include: financing bulk infrastructures (for example, roads, sewerage, water and electricity); training projects; business and entrepreneurial development within urban development programmes; and industrial development initiatives with other development agencies. Its annual investment in the provision of support is approximately R1 billion (HSA, 1991e).

The DBSA has prepared guidelines on the promotion and support of emerging construction industry entrepreneurs (Calcopietro, 1992). Calcopietro's (1992) summaries of (i) the model for DBSA involvement in the development of small-scale builders and (ii) the DBSA's recommendations for the provision of support services by institutions involved with the development of emerging contractors, are presented in Appendix D.

APPENDIX G

THE EVOLUTION OF SELECTED BLACK-OWNED SMALL-SCALE CONSTRUCTION ENTERPRISE IN SOUTH AFRICA

G.1   Source of Information

This section describes the evolution of a selected sample of certain successful black building contracting enterprises. This section of the literature review is based almost entirely on articles published during 1988, 1989 and 1990 in th journal ABC. Over this period the journal featured regular articles on black contracting firms and a column, "Succe Stories", compiled by news editor Dan Padi.

In total, forty-three black SSCEs (consisting of a mix of types of contractors) operating mainly in the PWV area we interviewed by Padi. The firms appear to have been discovered either through the various contractors' associations which they were members, or were found during random searches for activity on building sites. The contents of the articles appear to have been compiled from interviews held with the owners of contracting firms. Further, due to th similarity of the details given, it appears that the same questionnaire or semi-structured interviewing technique was used in gathering data from all of the firms. However, it cannot be said with certainty that the same set of question were put to each respondent. Clear patterns can be identified regarding the employment and training records of the firm owners, sizes and activities of the firms and opinions of the owners.

Many of the articles lack the depth of detail contained in others. In these cases it has been assumed that answers we not given by the respondents rather than that questions were not asked by the interviewer. Further, it must be noted that, although many interviewees reported having two or more partners, the interviews were conducted with only on representative of the firm in the majority of cases. Despite the fact that this may significantly bias the data, it was f that the articles were too rich a source of information to ignore and that their main value lies in the fact that the majority of them were compiled by the same author. The following review should, nevertheless, be treated with son circumspection.

The headings for sections G.2 to G.8 were derived by categorising the data in the articles. Thus, they are not necessarily exhaustive.

G.2   Educational and Training Background of Firm Owners

See Appendix A for details of education and training available for construction industry careers.

(a) Primary and secondary education : The articles reveal little detail about the schooling of the firm owners. Inde only three specifically stated that they had completed their secondary schooling (matric) (Padi, 1988a; ABC, 1988b; Padi, 1989c). One owner reported having left school after passing standard eight (Padi, 1990b) while another had abandoned his schooling after passing standard two. The latter is, however, literate and numerate and can read hous plans (Padi, 1989b).

It appears that the majority of firm owners did not complete their schooling. It is, however, unclear what the averag highest standard achieved was. Certain firm owners left school, but initially sought and obtained employment outsid the construction industry (ABC, 1988g; Padi, 1989b; Padi 1989c; Padi, 1990b, Padi 1990c).

There is little clarity on whether or not many firm owners had received technical secondary schooling. In fact only claimed to have done so. One, an electrician, had obtained the N1 (equivalent of standard eight) and N2 (equivalen standard nine) national technical certificates (NTC) (Padi, 1989c). Another firm owner reported that he held a NTC but gave no further details. A number of owners stated that they had "studied building" at "technical colleges". However, none of these individuals stated the exact qualification that they had obtained. It was therefore assumed th these individuals had failed to complete these courses, or alternatively, that the institutions were in fact trade school

(b) Tertiary education : None of the firm owners who were actually interviewed had obtained university degrees or

technikon diplomas. However, one interviewee claimed to have a partner who had qualified in civil engineering.

(c) Training : *(i) Training prior to starting enterprise* : Some firm owners stated that they had not obtained any formal building education at all before establishing their own businesses (ABC, 1988b; Padi, 1989a; Padi, 1989c; Padi 1990c). In this context, 'formal building education' refers to having obtained a qualification from an institution offering a course in building skills and/or theory. Indeed, many firm owners are likely to have avoided drawing attention to their lack of building education.

Relatively few (about one-third) of the firm owners underwent training or education at an institution (Padi, 1988a; ABC, 1988c; Padi, 1989a; Padi, 1989b; Padi, 1989c; Padi, 1990b; Padi, 1990c). A variety of descriptions were used to depict the type of institution attended, but the exact nature of the differences between institutions and courses cannot be established from the articles.

Only two firm owners specifically stated that they had attended BIFSA courses, one describing the course as a training course and the other as a management course (ABC, 1988g; Padi, 1990c). It is, however, possible that others attended BIFSA courses as some simply declared that they had attended trade schools, without naming the institutions.

Three firm owners reported having studied building immediately after leaving school. The institutions where they studied were "training colleges" in two cases (two years of study) and a trade school (three years of study) in the other case. Two of these firm owners appear to have established themselves in the building industry in a self-employed capacity immediately or very shortly after leaving these institutions. Significantly, one of these was the only firm owner who indicated that his firm was registered (see Appendix B about registration) with the Industrial Council and claimed to have built over five hundred houses and four blocks of flats over a period of fourteen years (Padi, 1988a; Padi, 1990b).

Notably, only one firm owner indicated that his employer had sent him to a "school of building" (Padi, 1989b). The remainder of those who had received a qualification in building appear to have done so on their own initiative (ABC, 1988c; Padi, 1989a; Padi, 1989c; Padi, 1990c; Padi, 1989b; Padi, 1989c; Padi, 1990b).

The vast majority of firm owners who received any form of education or training in building attended courses which focused on building skills or management. One firm owner attended a course at the University of Witwatersrand Business School. The course content was not given, but it seems probable that it was a short course in business management (ABC, 1988b).

Many firm owners gave no details of how they became skilled in their trades. It is likely that their employers trained them, but only four specifically indicated that this had been the case (Padi, 1988b; Padi, 1989a; Padi, 1989b; Padi, 1989c).

Five firm owners indicated that they had taught themselves their skills, mostly in their spare time. One of these had never before been involved in the building industry (ABC, 1988b; Padi, 1989b; Padi, 1989c; Padi, 1990c).

The articles reveal little evidence that firm owners were operating or specialising in trades other than those in which they had been trained. Only two clear references are made to such trade shifts, a painter who became a builder (presumably bricklaying) (Padi, 1990b) and a plumber who, after twenty four years, started building (Padi, 1990b).

(ii) Training after starting enterprise : Evidence of firm owners having pursued further skills training or building education was almost entirely absent in the articles. Indeed, only two firm owners indicated that they had received training since starting their businesses. Significantly, in neither case was this a development of their usual skills, but was rather an attempt to broaden their skills bases. One took an "electrical wiring" course for two years, believing that it would enhance his building knowledge and chances of success (Padi, 1988a). The other firm owner was enroled on a "TV technician" course for unspecified reasons (Padi, 1988a). Both of these endeavours appear to be attempts at providing employment assurance engendered by the perceived instability of the house building industry.

A few firm owners claimed that they provided skills training for their own employees and in all cases this involved bricklaying (Padi, 1989b; Padi, 1990c). Although no details were given, it seems unlikely that this training would have involved any theoretical component.

(iii) Training intentions : Only one firm owner stated that he planned to further his education and intended to enrol a "management course". He himself had received BIFSA training and also intended sending his employees on BIFS courses (ABC, 1988g).

(d) Summary: Firm owners had generally not completed their secondary schooling. Further, the majority had not received any technical schooling or tertiary education. It was also found that formal trade skills training qualificatic were uncommon.

G.3   Entry Into the Building Industry

(a) Reason for entering the building industry : The activities which a general worker may legally perform are documented in great detail in the various Industrial Council agreements (see for example Republic of South Africa, 1987a and 1987b). Common to all of these permissible activities is their unskilled nature. It is therefore likely that many building industry employees originally entered the industry because work was available which either did not require or expect any particular qualification or training. Most of the firm owners interviewed by Padi, stated that e "entered the industry" on a given date, but gave no reason why this was their choice. It is possible that many enter unskilled positions in with the intention of gaining access to on-the-job training.

Those firm owners who did mention their motive for entry into the industry had done so for positive reasons. Some these owners had been aspirant builders since their childhood or school days (Padi, 1988a; ABC, 1988g; Padi, 198 Padi, 1989c; Padi, 1990b). Notably, they had entered the industry in the pursuance of planned career objectives.

The housing backlog was another factor which attracted some firm owners to the industry. This attraction was base on the belief that a steady supply of work would be generated by the need for housing (Padi, 1988a; Padi, 1990a; Padi, 1990c).

(b) Status on entering the building industry : Approximately one quarter of the firm owners gave no indication of their positions on first entering the building industry. However, most firm owners did furnish these details, or it wa reasonably clear from the articles what their positions were. The majority of firm owners started in the industry as general workers (unskilled employees) (Padi, 1988b; Padi 1989a; Padi, 1989b; Padi, 1989c; Padi, 1990c). Approximately one fifth of all owners, however, entered the industry as skills trainees in the employ of various construction companies. Notably, only half of these individuals later established firms specialising in their particula skills (Padi, 1989c; Padi, 1990a; Padi 1990b; Padi 1990c).

Only six individuals claimed to have entered the industry as owners of, or partners in, independent contracting or s contracting enterprises (Padi, 1988a; ABC, 1988g; ABC, 1988b; Padi, 1990a).

(c) Summary: Only a minority of firm owners gave reasons as to why they had entered the building industry. Thus general trend regarding motive for entry could not be established. Those that did give reasons, had entered for posi reasons - mainly because they had planned careers in construction, or wished to help reduce the housing backlog.

G.4   Establishment of Firms

(a) Reasons for establishment : Ambition, founded on a desire for independence or discontentment with employee status, was given by some firm owners as the reason for starting their own firms (Padi, 1988a; Padi, 1988b; Padi, 1990b). One case involved an employee technician in the manufacturing industry who went into marketing, then established a building material supply business and finally started a building construction firm (Padi, 1988a).

In one case a firm owner was motivated to start his own business because he felt he had been exploited by his employer, a large white-owned contracting company. This allegation arose out of the fact that although he was doir skilled work, he was paid less than the other skilled workers. His argument was that, despite his lack of qualificatie or formal training, his employer found his work to be satisfactory and therefore he believed it merited the full wage (Padi, 1989c). The issue of whether or not exploitation occurred is less important than the resultant action of the individual. The perception that an injustice had been committed was the catalyst for the emergence of entrepreneurial behaviour.

Retrenchment is common in the building industry and frequently workers are laid off by a company on completion of a project, but soon find employment elsewhere. However, only one firm owner claimed to have established his business as a result of being retrenched by a large white-owned construction company (Padi, 1990b). It is possible that this was, indeed, the case for many other firm owners, but it is reasonable to expect that this information may have been withheld by individuals wishing to avoid creating the impression that they lacked competence.

(b) How firms were established: A distinction must be made between: those owners who had previously been employees; those who had previously owned firms other than building contracting businesses; and those who had expanded their previous building contracting firms.

The majority of firms appear to have been started by individuals who had previously been employees of large contracting companies. No detail regarding exactly how these firms were originally organised is given in the articles. It is likely though, that the majority of these firms were founded haphazardly, with little or no capital and perhaps a few small contracts lined up. The general pattern seems to be that the owner possessed a skill, employed some assistants and took on whatever work was available.

However, five firms were established as a result of the fragmentation or amalgamation of existing businesses. Four of these cases involved the establishment of new firms while the remaining firm expanded (ABC, 1988d; Padi, 1989a; Padi 1990b; Padi 1990c; Alexander, 1990). These firms largely regarded the resultant broader base of expertise as the key to their success. An important factor which some firm owners emphasised was that their previous businesses still existed and that the new ventures had not been established as a result of the failure of previous ventures.

(c) Summary: Common reasons for having established building businesses included: ambition; desire for independence; and discontentment with employee status. Relatively uncommon reasons included perceived exploitation by previous employer and retrenchment. The majority of firms were established by individuals who had previously been employees of larger contractors - probably once the individual had acquired what he perceived to be a sufficiently sound skills base to justify undertaking his own contracts. It was relatively uncommon for firms to have been established by individuals who had not been construction industry employees, or by individuals who had previously owned contracting businesses.

G.5   Track Record of Firms

(a) Nature of first contract and how acquired : Many firm owners could not remember, or failed to state, precisely what work they first took on. Half of the interviewees did, however, explain the nature of their firms' first project. These projects can be categorised as sub-contracts, independent contracts and partially independent contracts.

The majority (twelve) of these firms started out doing sub-contracts for developers or construction companies (Padi, 1988a; Padi, 1989b; Padi, 1989c; Padi, 1990b; Padi, 1990c). None of these firms, however, specifically stated how the sub-contracting work was acquired.

Eight firms began operations doing **independent contracts**, apparently having sought and acquired work without relying on larger contractors (Padi, 1988a; Padi, 1988b; ABC, 1988b; ABC, 1988g; Padi, 1989a; Padi, 1989b; Padi, 1990a; Alexander, 1990). Although not stated, it is unlikely that these firms would have formally advertised their services, but probably became known to private individuals as independent contractors. The majority of these contractors reported having done mainly alterations jobs.

Only one firm (ABC, 1988b) reported having been **partially dependent** in that its first project was for Home Development Company (HODECO).

(b) Nature of subsequent work and shifts in type of work : Some important facts emerged from an analysis of the subsequent activities of the firms dealt with in (a) above. All except one of those who had started as sub-contractors,

remained sub-contractors. The firm that moved away from sub-contracting went on to establish itself as an indeper contractor securing its own contracts for new work (Padi, 1989c).

All of the firms who had started as independent contractors subsequently changed the nature of their activities. On these firms entered the sub-contracting market (ABC, 1988c). The most significant trend, however, was that all of firms which had started out doing alterations progressed and began acquiring new buildings (Padi, 1988a; ABC, 1988g; Padi, 1989a; Padi, 1989b; Padi, 1990a). One firm which had initially done alterations work became involv in building commercial buildings (Padi, 1988b).

The firm which had commenced business working under HODECO progressed and began financing its own projec and acquiring development sites through a local contractors' association (ABC, 1988b).

The majority of **firms which had not given details of their first contracts** did, however, report on the type of w they had done subsequently. The types of work that these firms had been involved in concur with the categories defined in (a) above.

The majority of firms (eleven) in this group had acquired **independent contracts** since starting up (ABC, 1988d; 1989a; Padi, 1990a; Padi, 1990b; Padi, 1990c).

This group of firms included a number who had been involved in **partially dependent contracts** working for, or the following organisations: HODECO (one firm) (ABC, 1988f); South African Housing Trust (SAHT) (one firm) (Padi, 1990c); both HODECO and SAHT (one firm) (Padi, 1990b); and HODECO and the Small Business Development Corporation (SBDC) (one firm) (Padi, 1989a).

Only one of the firms in this group had been doing sub-contracts.

(c) Rates of production : Details were given in (a) and (b) above regarding the type of work that certain firms hac started out doing and what they had done subsequently. This section reviews the amount of work done since the establishment of the firms. In this section, firms are grouped according to the nature of their most recent principal of work. Unfortunately, the information on ten of the firms was insufficient for the purpose of establishing their re of production.

Most of the sub-contracting firms reported extensively on the number of houses or buildings that they had built si commencing business. However, due to the fact that a principal contractor would originally have acquired the wor these numbers could at best only indicate how busy principal contractors had been. Details of rates of production these firms have consequently been omitted.

It was possible to calculate rates of production for thirteen **independent contractors**. Nine of these firms produce average of fewer than twelve houses per annum (Padi, 1989a; Padi, 1989b; Padi, 1990b; Padi, 1990c). The remai four firms respectively produced an average of: 13 houses p.a. (Padi, 1988a); 17 houses p.a. (Padi, 1990a); 35 ho p.a. (ABC, 1988d); and 67 houses and 2,5 shopping centres p.a. (Padi, 1988b).

Only two of the **partially dependent** contractors gave sufficient details for the calculation of their rates of product In one case an average of 0,75 houses p.a. were reported, but it seems highly probable that the owner neglected tc mention some of his completed projects (Padi, 1989a). The other firm produced an average of 10 houses p.a. for HODECO and SAHT collectively and an average of 2 houses p.a. on his own finance (ABC, 1988b).

(d) Future work : Six firm owners gave details of future projects that had been secured. This involved fewer than houses in three of these cases (Padi, 1988a; Padi, 1990b). However, the three remaining firms respectively reporte future commitments involving the development of: 300 sites (Padi, 1990a); 2200 stands (in a consortium of 7 companies) (ABC, 1988d); and a multi-purpose centre valued at R3 million (Padi, 1988a).

(e) Sources of finance : Unfortunately the articles revealed very little data about the firms' sources of finance. However, two firm owners indicated that their initial source of finance had been small amounts of their own savin One firm started on R600 (in 1982) (Padi, 1988b) and the other commenced business with R2 000 (in 1986)

(Alexander, 1990). Two firms reported that they had initially obtained finance from the SBDC (ABC, 1988b; Padi, 1990c).

Even less indication was given by the firms regarding subsequent or regular sources of finance. One firm admitted to having experienced difficulty in obtaining finance (Padi, 1988a), another identified credit from its materials supplier as its main source of finance (Alexander, 1990), while another suggested that its contractors' association provided bridging finance in times of crisis (Padi, 1990c). Two firms specifically stated that loans from commercial banks and formal lending institutions were unattainable to themselves, and, indeed, to most black SSCEs (Padi, 1990c; Alexander, 1990).

(f) Summary: Common sources of first contract was sub-contracting to a larger contractor or personally obtaining work from individuals. Notably, it was extremely uncommon for firms to have obtained their first contract through institutions such as those described in Chapter 4. Sub-contractors tended to continue sub-contracting over the life of the firm, but independent contractors generally progressed from initially doing small alterations jobs to building new structures. The majority of firms had produced approximately one house per month, although there were reports of individuals completing up to 6 houses per month. Only a minority of firm owners appeared to have secured work for completion at a future date. Little data was available on initial and subsequent sources of finance, but it seems likely that the owners' own savings had been a common source.

G.6   Life Span of Firms

It was possible to identify the life span of twenty four firms. These life spans were all calculated by taking the difference between the year of commencing business and the publication of the article.

The majority of these firms had been in existence for less than nine years. This group consisted of eight firms with life spans of three years or less (ABC, 1988d; Padi, 1989a; Padi 1989c; Padi, 1990a; Padi, 1990b; Padi, 1990c) and nine firms with life spans greater than three years, but less than nine years (Padi, 1988a; Padi, 1988b; Padi, 1989a; Padi, 1989b; Padi, 1990b; Alexander, 1990).

A second group of firms was identified that had been in existence for longer than eight years, but less than fifteen years. Five firms fell into this category with life spans of: ten years (one firm - Padi, 1989a); twelve years (one firm - Padi, 1989a); and fourteen years (three firms - Padi, 1988a; Padi, 1989b; Padi, 1990c). The third group consisted of two firms with life-spans of eighteen years (Padi, 1990b) and twenty six years (Padi, 1990c) respectively.

Since the entire sample consisted of successful contractors, these life spans should not be considered as being representative of all SSCEs. It is likely that the life-span of the average SSCE would be relatively shorter than these data suggest.

G.7   Employment

(a) Types of permanent staff : Various types of permanent staff employed by the firms were identified namely, casuals, assistants (presumably unskilled general workers), artisans (presumably both qualified and unqualified individuals performing skilled work), sub-contractors and clerical or administrative personnel. In the case of ten firms, however, the owners themselves (usually artisan bricklayers) performed the skilled work and employed assistants only (Padi, 1988b; Padi, 1989a; Padi, 1989b; Padi, 1989c; Padi, 1990b; Padi, 1990c).

Eleven firms gave sufficient detail to identify the ratio of skilled to unskilled employees. Four firms worked on a ratio of 1 labourer to 1 artisan (Padi, 1988a; ABC, 1988b; ABC, 1988g; Padi, 1990b). Six firms, however, employed 2 labourers per artisan (ABC, 1988f; Padi, 1989a; Padi, 1990a; Padi, 1990b). Only one firm was found to employ 3 labourers per artisan (Padi, 1990c).

Four firms recorded that they either supplemented their permanent work force with sub-contract labour (ABC, 1988b; ABC, 1988g) or that they only ever employed sub-contractors (Padi, 1989a; Padi, 1990b).

Only two firms reported having permanent clerical or office-bound staff. Both firms claimed to employ "architects" (Padi, 1990b). *These are unlikely to be registered architects (author's note).*

(b) Total number of permanent employees : Total numbers of permanent staff were calculated for twenty two firms. Four of these firms employed fewer than 5 people on a full-time basis (Padi, 1988a; ABC, 1988b; Padi, 1989b; Padi, 1990b). The majority (thirteen firms), however, employed more than 4, but fewer than 13 permanent staff members (Padi, 1988a; ABC, 1988f; ABC, 1988g; Padi, 1989a; Padi, 1989b; Padi, 1990b; Padi, 1990c). Three firms reported considerably larger numbers - 24 employees (one firm - Padi, 1990a), 42 employees (one firm - Padi, 1990a) and 15 employees (one firm - Padi, 1988b). It seems highly unlikely that the latter figure was representative of the firm's usual full-time staff. It probably indicated the total number of individuals that were involved on the large project (see section G.5 (d) above) it was controlling at the time of the interview.

(c) Turnover of employees : Although it was not clear whether or not the firms had generally experienced a low or high turnover of staff, the following may be of some interest.

Only one firm specifically mentioned that it had regularly used casual labour to supplement its work force (Padi, 1990c). It is, however, likely that this was the case for many other firms.

Six firms claimed to have kept the same employees since commencement of operations. Two of these had, however, not specified their life spans (Padi, 1988a; Padi, 1990b). The remaining four firms had operated with the same staff respectively for periods of: 3 years (one firm - Padi, 1990a); 4 years (two firms - Padi, 1990b; Padi, 1990c); and 10 years (one firm - Padi, 1989a).

A further three firms had kept the same staff together for reasonably long periods of time - 8 years (Padi, 1989a), 10 years (Padi, 1988a) and 15 years (ABC, 1988c).

(d) Training of employees : Four firm owners, all bricklayers, reported that they were teaching their assistants the bricklaying trade (Padi, 1989b; Padi, 1990c). Further, one firm owner reported that he had taught two employees "to build" and that both had gone on to establish their own building enterprises (ABC, 1988c). In another case the firm owner reported that some of his former employees had established their own building companies, although he did not claim responsibility for training them (Padi, 1989a).

(e) Summary: Many firms appeared to rely on the owner as the sole source of artisan skills and it also appeared uncommon for firms to employ support staff such as clerks, secretaries, *etc.*. The majority of firms had fewer than 13 full-time employees and it appeared uncommon for firms to experience a high turn-over of permanent employees. Although not widespread, there was evidence that some firm owners had employed unskilled individuals, with the intention of developing their skills through on-the-job training provided by the owner himself.

G.8   Resources

(a) Business premises : There is no evidence in the articles suggesting that any firms actually owned their business premises. Notably, ten firms were run from the owners' homes, although this was probably also the case for the vast majority (Padi, 1988a; Padi, 1989a; Padi, 1989c; Padi, 1990b; Padi, 1990c; Alexander, 1990).

(b) Equipment : The only indication in the articles that firms owned major pieces of plant or equipment involved vehicles. One firm reported that its assets included 1 car, 3 vans and 10 trucks (Padi, 1988b), while another made less specific reference to its "trucks and cars" (Padi, 1990b).